VEHICLE MAINTENANCE AND REPAIR

LEVEL 1

Patrick Hamilton, John Rooke and Julian Brown

Edited by Roy Brooks

DELMAR
CENGAGE Learning·

Australia · Brazil · Japan · Korea · Mexico · Singapore · Spain · United Kingdom · United States

DELMAR
CENGAGE Learning·

Vehicle Maintenance and Repair Level 1
Patrick Hamilton, John Rooke and Julian Brown
Edited by Roy Brooks

Publishing Director: Linden Harris

Commissioning Editor: Lucy Mills

Editorial Assistant: Claire Napoli

Project Editor: Alison Cooke

Production Controller: Eyvett Davis

Marketing Manager: Lauren Mottram

Typesetter: MPS Limited

Cover design: HCT Creative

Text design: Design Deluxe, Bath

For product information and technology assistance, contact **emea.info@cengage.com**.

For permission to use material from this text or product, and for permission queries, email **emea.permissions@cengage.com**.

British Library Cataloguing-in-Publication Data
A catalogue record for this book is available from the British Library.

ISBN 13: 978-1-4080-6422-1

Cengage Learning EMEA
Cheriton House, North Way,
Andover, Hampshire SP10 5BE
United Kingdom

Cengage Learning products are represented in Canada by Nelson Education Ltd.

For your lifelong learning solutions, visit **www.cengage.co.uk**

Purchase your next print book, e-book or e-chapter at **www.cengagebrain.com**

Printed in the United Kingdom by Ashford Colour Press Ltd
Print Number: 11   Print Year: 2023

MIX
Paper from
responsible sources
FSC® C011748

I would like to thank my wife Sharon, daughters Jessica and Kirsten, also my late stepson Peter and stepson Simon, for their patience and support whilst I have worked on this book

Patrick Hamilton

I would like to thank Jackie for supporting and encouraging me during the many hours writing the workbook

John Rooke

I would like to take this opportunity to thank everyone who has assisted me throughout my career. I would also like to particularly thank my children who continue to make me a very proud father

Julian Brown

CONTENTS

FOREWORD

Welcome to this very latest edition in the Vehicle Maintenance and Repair series, popularly known as the "Brooks Books". For more than 40 years these books have helped many tens of thousands of motor vehicle students to gain and conveniently record knowledge in the exciting world of automobile engineering.

Although the basic essentials, such as "suck, squeeze, bang, blow" must remain the same, automotive technology and the way in which it is taught is constantly evolving. Similarly so with the books which from the very beginning have been frequently subject to technical revision and variation of content to suit ever changing needs.

This latest full colour edition with a high proportion of specially drawn illustrations, maintains the original ease of use and understanding, but adds considerable value by including helpful practical tips, guidance to useful websites, word puzzles and the like – all designed to stimulate interest and remain a valuable source of reference. With due deference to the long-established highly successful format of the books, the new team of authors, all practising and experienced lecturers in motor vehicle work, clearly show that they understand and accurately cater for the needs of both students and teachers.

Roy Brooks, Editor

ACKNOWLEDGEMENTS

The publisher wishes to thank the following companies for granting permission to use images, drawings and artworks:

BMW	Ford	Snap-on Tools Ltd.
Bosch	Honda	Volkeswagen
Bridgestone	Mercedes	Vauxhall
Comma oil	Peugeot	
Draper Tools	Sealey	

Photo research and permissions by Alex Goldberg, Jason Newman and Tara Roberts of www.mediaselectors.com.

Thank you to Daniel Carey and the Honda Institute for their kind help sourcing images.

ABOUT THE AUTHORS

Patrick Hamilton CertEd PGDE LCGI EngTech MIMechE AAE MIMI MIfL

Patrick has over 20 years light and heavy vehicle practical experience gained at leading vehicle manufacturers' main dealerships, including 6 years in the Royal Air Mechanical Transport Servicing Section. He is currently Head of the School of Engineering at West Suffolk College and teaches on automotive and engineering courses. Patrick has experience of teaching on a range of light vehicle and heavy vehicle programmes ranging from entry level to Level 4. He has also worked for the Sector Skills Council writing some of their QCF units and has been a technical consultant and author for City and Guilds and the IMI Awards, helping to develop and write their qualifications.

John Rooke CertEd, MIMI, MifI

John Rooke has over 30 years experience in the motor trade trained at a main dealer and he then took the opportunity to enter teaching as an instructor in a private training school, teaching practical skills. In 1999 he became a lecturer at Cambridge Regional College teaching Automotive Engineering to learners from entry level to Level 3. John presently has an additional role of supporting engineering staff to improve teaching and learning. During the past 9 years he has been the college's IMI centre coordinator enabling him to keep fully up to date with qualification and assessment requirements.

Julian Brown BA (Hons) QTLS CertEd MifI

Julian has served in the motor vehicle industry for over 30 years, starting out as a HGV/PSV apprentice at a local haulage company. For 10 years he ran his own business undertaking all aspects of repair and maintenance specializing in plant and agricultural equipment. Eight years ago he had the opportunity to join Stoke-on-Trent College as a lecturer where he has achieved all of his teaching qualifications, assessors, verifiers, and more recently QTLS and a BA (Hons) in Post Compulsory Education. As one of the course team leaders for the apprentice team he is currently responsible for over 100 HGV, LV and B/P apprentices. He also has worked with the IMI and was lead author for the City and Guilds on the development of framework one.

QUALIFICATION MAPPING GRID

The following grid shows some of the units from awarding bodies which this book either partially, or completely, map across to. It is designed to aid those delivering the qualifications to be able to see at a glance.

The Foundation Learning units have no generic QCF numbers, unlike the level 2 and 3 qualifications, that form a part of the apprenticeship frameworks. For this reason either the awarding body unit number, or in the case of ABC Awards, their own QCF unit number has been used.

	IMIAL	ABC Awards	City & Guilds
Personal and social development	PW1 DWP1 DS1		
Employment rights and responsibilities for the automotive sector (ERR)	ERR1		
Health and safety	EL01 L101 HV1	H/501/7005	001 051
Tools, equipment and materials	L102 HV2	A/501/7009	026/076 712
The retail automotive industry	EL02	D/501/7021	031 81
Engine construction and operating principles	L103 L104 L105 L110 L112 L119 L120 HV5	M/501/7010 T/501/7011 A/501/7012 F/501/7013 J/501/7014 M/501/7024 A/501/7026	102 152 172 701 702 703 704
Chassis systems	L106 L107 L109 HV3 HV4	R/501/7016 Y/501/7017 D/501/7021	154 551 553 706 708 709
Driveline	L111 HV6	J/501/7014	112 162 707
Electrical systems	L113 L114 HV2 HV7	L/501/7015	103 153 705
Low carbon technologies	EL121 EA2		
Valeting	VV01K VV02K	J/501/7028	
Heavy vehicle maintenance	HV3 HV4		
Motorcycle maintenance	EL14 L115 L116	L/501/70129	

Note: Aspects of optional units are contained within relative sections of the workbook. It should be remembered that whilst the main learning objectives of the qualification have been included, it is important to refer to the latest QCF structure and units to ensure full coverage.

ABOUT THE BOOK

With the help of your supervisor, try using this equipment to lift an engine and transport it to another location in the workshop.

Activity boxes provide additional tasks for you to try out

http://www.direct.gov.uk

Web link boxes suggest websites to further research and understanding of a topic

If the load is awkward ask someone to assist the lift. With heavy loads, use lifting equipment.

Tip boxes share author's experience in the automotive industry, with helpful suggestions for how you can improve your skills

If the jack is faulty, do not use it and report the fault to your workshop supervisor.

Health and safety tip boxes draw your attention to important health and safety information

When working on engines (as with all vehicle systems) refer to the vehicle manufacturers' repair instructions for the torque settings, fluid types and quantities.

Routine maintenance procedure boxes provide advice and guidance when carrying out routine maintenance tasks on a vehicle

Online lecturer resource Check answers to all of the student activity questions in this book. Please register here for free access: http://login.cengage.com

Multiple choice questions

Choose the correct answer from a), b) or c) and place a tick [✓] after your answer.

1 **The correct way to finish writing an official letter is with:**

   a) Yours Kindly [ ]

   b) Yours Faithfully [ ]

   c) Best Regards [ ]

2 **'SMART' is an abbreviation for:**

   a) Specific, Measurable, Achievable, Reliable, Time-bound [ ]

   b) Specific, Measurable, Accountable, Realistic, Time-bound [ ]

   c) Specific, Measurable, Achievable, Realistic, Time-bound [ ]

3 **How many pages are recommended for a CV?**

   a) One or two [ ]

   b) Two or three [ ]

   c) Three or four [ ]

Multiple choice questions are provided at the end of each chapter. You can use questions to test your learning and prepare for assessments

Learning objectives

After studying this section you should be able to:

● **Identify the main motorcycle and moped components that require maintenance**

● **Describe the basic motorcycle maintenance procedures**

Learning objectives at the start of each chapter explain key skills and knowledge you need to understand by the end of the chapter

A QUICK REFERENCE GUIDE TO THE QUALIFICATION

Traditionally, learners would have attended college full time to achieve a Vocationally Related Qualification (VRQ) or part time as part of an apprenticeship, gaining a National Vocational Qualification (NVQ) and a VRQ as well as the required key skills. To an extent this has not really changed.

What has changed is how the qualifications have been developed, allowing a more flexible approach to the selection of units and their delivery. This has been developed by the Sector Skills Council (SSC) for the industry area concerned. The SSC for the automotive industry is known as Automotive Skills, a division of the Institute of the Motor Industry (IMI). The SSC is employer led and acts in response to employer needs. It does this by producing National Occupational Standards (NOS), developed by employer partnerships and working parties. The NOS describe the different functions carried out by people working throughout the range of sectors in the industry. The Skills, Knowledge and Competency requirements are identified for all the levels of technical, parts, sales and operations management. Qualification Credit Framework (QCF) units have been developed from the NOS. These are the common units which awarding bodies, such as IMI Awards, City and Guilds and Edexcel Btec use to develop their qualifications. This creates a method of standardization across all awarding bodies.

One of the advantages of the QCF units is that a learner can complete individual units at one training provider using a specific awarding body. They may, for whatever reason, have to move to another training provider who uses a different awarding body, whereby the accredited units are transferable.

The QCF units cover Competency, Skills and Knowledge. Each unit is allocated a "Credit" value. A predetermined number of "Credits" are required to achieve a qualification. The QCF units and "Credits" are transferable across awarding bodies, allowing the learner to build and complete their VRQ and/or Vocational Competence Qualification (VCQ) even if they move around the country.

The Knowledge units cover the technical understanding of the subject area. The Skills units show that the learner is able to carry out practical tasks to a required standard.

These units are designed to be delivered in a college and training provider environment. The Competency units show that a learner not only has the required skills but that they are now able to perform tasks independently within given timescales and with limited support and guidance. The Competency units can only be completed in the workplace.

The VRQ is specifically designed for a training environment based delivery on either a full or part-time basis. With the use of the QCF units it is divided into two areas, those of Skills and Knowledge. A variety of qualifications have been designed to meet the VRQ criteria which are constructed with a mixture of the QCF units. These can provide learners with practical skills and knowledge preparing them for work in the automotive industry.

The Vocational Competence Qualification (VCQ) (originally known as the NVQ) covers the Competency requirements needed for an individual to satisfactorily perform and function, for example, at Level 2.

Employers can select QCF units which meet their business and strategic needs.

This book has been designed to assist the learner in developing their knowledge and skills to at the Foundation Learning level. It covers a majority of the requirements to meet the light vehicle, heavy vehicle and motorcycle requirements at these levels, for IMIAL, City and Guilds, Edexcel Btec and ABC Awards.

FOUNDATION LEARNING

This is the name given to the Entry Level 3 and Level 1 education provision. These form a part of the wider 14-19 and vocational qualification (VQ) reform programme. It aims to improve the skills of learners by helping them gain credit-based qualifications at Entry Level and Level 1 in the Qualifications and Credit Framework (QCF), which can then lead them to appropriate destinations such as Level 2 qualifications, supported employment or independent living.

PART 1
INTRODUCTION

USE THIS SPACE FOR LEARNER NOTES

_____

_____

_____

_____

_____

_____

_____

SECTION 4
Tools, equipment and materials 30

SECTION 5
Introduction to the retail automotive maintenance and repair industry 51

SECTION 1

Personal and social development

USE THIS SPACE FOR LEARNER NOTES

_____

_____

_____

_____

_____

Learning objectives

After studying this section you should be able to:

● Identify and tackle straightforward problems.
● Plan for self-development.
● Prepare for work.

Key terms

Curriculum Vitae A document containing details of someone's course of life and achievements.
Teamwork Cooperating with and caring about other workers.
Communication Listening, reading, speaking and writing.

www.citizensadvice.org.uk

www.bbc.co.uk/skillswise

www.workteams.org

INTRODUCTION

To be able to work well and progress in life, we all need to have a set of personal skills, like a tool box which can help us develop and reach our full potential. These personal and life skills can include:

- **Teamworking**
- **Communication**
- **Problem solving**
- **Preparation for work**
- **Healthy living and eating**
- **Environmental awareness**

DEALING WITH PROBLEMS IN DAILY LIFE

Working in the motor vehicle industry, like any other industry, requires us to be able to deal with problems which could be at work, home and in the world at large.

Every day we encounter challenges. How we approach and deal with these challenges is very important and can affect the outcome of these challenges.

Three types of problems which are likely to be encountered are:

Personal

Give ONE example: _____

Financial

Give ONE example: _____

Technical

Give ONE example: _____

When a problem or activity comes our way we need to be able to work through it using a systematic approach. This means doing things one at a time and in a suitable order.

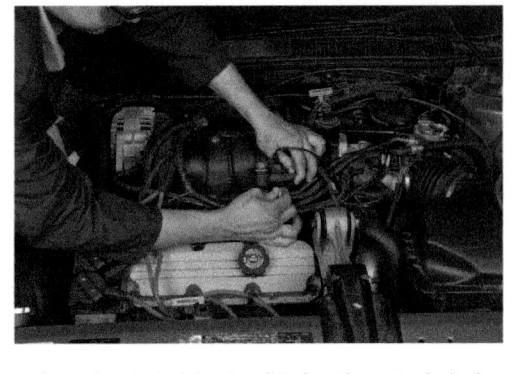

A service technician troubleshooting a technical problem

Complete the following table:

Problem	Financial, personal, social or technical	How can this problem be approached/tackled?	Ways to check if the problem has been solved
A customer's car will not start	_____	_____	_____
		_____	_____
Arguing with colleagues	_____	_____	_____
		_____	_____
		_____	_____
Not being able to afford a new socket set	_____	_____	_____
		_____	_____
		_____	_____
		_____	

DEVELOPING SELF

We all have personal strengths and abilities. These help us to do well and succeed in life.

These can include:

- **Enthusiasm**
- **Physical strength**
- **Communication skills**

Name THREE more:

_____

_____

_____

There are other personal strengths and abilities. We normally have some of these, and some people may have all of them. Make a list of your personal strengths and abilities.

Let us look at one area which may need developing. It could be subject knowledge. You may not be very good at electrics and know that to progress as a technician you need to improve this subject knowledge.

A way to achieve this goal is by using SMART targets.

Complete the following:

S_____ = What exactly is it that needs to be achieved? (e.g. complete the 10000 mile service

on the car by 5 pm)

M_____ = How do I know when this is achieved? (e.g. is the service done by 5 pm?)

A_____ = Can I do it? (e.g. do I have the parts and tools?)

R_____ = Is it achievable? (e.g. what is the manufacturer's time for the job?)

T_____ = How long is needed to achieve it? (e.g. it must be done by 5 pm?)

 Think of an area in your life which you need to develop and use SMART targets to work out how you can achieve this.

PREPARATION FOR WORK

The time will come when you will need to apply for a job. It is very likely that a lot of people will be applying for the same job, so it is very important you are able to 'sell' yourself as being the best person for the position.

There are a number of job roles in the motor industry, some of which may be of interest to you. You will need to look at your qualifications, experience and skills set to determine if you are suited to the job being advertised.

If you do not have the skills, experience and qualifications for a job that you would like to apply for, what can you do to gain them?

Also, and probably most important, is that you are keen and enthusiastic about this job role. Is it a job you really want to do?

List FIVE likely places or methods for advertising a job:

1 _____

2 _____

3 _____

4 _____

5 _____

After seeing a job advertised the next step is to apply for it. The advert may ask you to telephone for an application form, or may ask for a covering letter and a CV.

What is CV an abbreviation of? _____

What does this mean? _____

Ideally, how many pages should a CV have?

_____

There are a number of different formats for a CV, but they will generally all contain the same relevant information.

In the CV heading you can write your general information:

- **Name**
- **Surname**
- **Address**
- **E-mail address**
- **Telephone number**

List the main information that a CV will need to contain:

_____

_____

_____

_____

62170 Help Wanted-Skilled Trades

ASPHALT
Experienced equipment operators, foremen and laborers. Good pay and benefits. Apply in person

Auto
A+ TECHS
General Motors Line Techs needed for Chesrown Pontiac-Buick-GMC in Delaware. Call Gerry or Frank

Automotive
LINE TECHS
For Chesrown Service Center at Karl & Morse Rd. Top Wages! Electrical and a/c a must. Call Jay

AUTOMOTIVE TECH
Fully tooled, specializing in engine removal and intsallation. Competitive pay, weekends off, uniforms. Call Randy:

BRUNSWICK MECHANIC
time-Shifts vary, experience but will train. Apply in

Check your local newspaper for businesses that are looking for apprentices or technicians

When applying for a job, an application form may need to be completed and sent to the potential employer. This can be done either on paper or submitted electronically. It is more than likely that you will be asked to send in a copy of your CV and a covering letter.

What information is needed in a covering letter?

_____

_____

_____

Always remember to be polite and courteous in your letter writing. A good way to end a covering letter is:

'Thank you for your time and I look forward to hearing from you in the near future.

Yours faithfully

Your signature

Your printed name'.

When you are successful with an application and have been invited in for interview, you will need to plan and prepare for this. Before attending the interview what are the important things which must be planned for?

_____

_____

List FIVE personal qualities that an employer is likely to want from an employee:

1 _____

2 _____

3 _____

4 _____

5 _____

Using the information covered in this section, write your CV. Keep in mind the type of job you might be applying for and highlight the areas of your life which will be useful for this job.

Communication

In many instances, problems and misunderstandings arise as a result of inadequate communication.

An essential factor contributing to the smooth efficient operation of an organization is the establishment of clearly defined *lines of communication.*

State the methods of communication employed within an organization when dealing with customers and colleagues:

Good customer relations are important; make sure you always listen and communicate clearly

_____

Complete the following table that shows examples of communication:

Examples of communication	Preferred method of communication
Consulting with colleagues and customers.	_____
Quick direct communication with colleagues and customers at a distance.	_____
Preparation of job sheets.	_____
	_____
Quick easy communication with colleagues and customers at a distance.	_____
Recording customer and vehicle details.	_____
	_____
Sending reports to customers and colleagues, communications.	_____
Reports and letters.	_____

Positive working relationships develop with the people you interact with at work. List some people with whom you will need to maintain a positive working relationship.

_____

_____

Good working relationships are very important to the success of every business.

Teamwork

Working as a team

Insert these missing words into the following paragraphs (note: there are two extra distracter words):

co-operate	employees	less	shrink
hard	teamwork	recommend	efficiently
members	productive	efficient	reputation
enthusiastic	workplace	resources	
good	image	morale	
trust	more	grow	

To make the company successful, all of its _____ must work together. They must

_____ and care about other workers; like _____ of a football team: this is _____.

If the company does well, and employees get on with each other and _____ each other,

there will be a _____ feeling in the _____, and people will be _____ about their

jobs. This feeling is called good _____.

When everyone works _____, and no one wastes time or _____, the company will be

_____. By being efficient, employees will get a lot done – they will be _____.

If a team with _____ morale works _____ and productively, customers will be satisfied

and pleased to come again. They are also likely to _____ the company to others, so it will

gain a good _____. This in turn will bring more business, and the company will become

even _____ successful, and will _____. It will gain a good company _____.

Building good relationships

List SEVEN things which contribute to good working relationships:

1 _____

2 _____

3 _____

4 _____

5 _____

6 _____

7 _____

List some of the things which could upset good working relationships:

● _____

● _____

● _____

● _____

● _____

● _____

Multiple choice questions

Choose the correct answer from a), b) or c) and place a tick [✓] after your answer.

1 **The correct way to finish writing an official letter is with:**

a) Yours kindly [　]

b) Yours faithfully [　]

c) Best regards [　]

2 **'SMART' is an abbreviation for:**

a) Specific, Measurable, Achievable, Reliable, Time-bound [　]

b) Specific, Measurable, Accountable, Realistic, Time-bound [　]

c) Specific, Measurable, Achievable, Realistic, Time-bound [　]

3 **How many pages are recommended for a CV?**

a) One or two [　]

b) Two or three [　]

c) Three or four [　]

SECTION 2

Employment rights and responsibilities for the automotive sector (ERR)

USE THIS SPACE FOR LEARNER NOTES

_____

_____

After studying this section you should be able to:

- Determine minimum wage requirements.
- State the legal requirements relating to contracts of employment.
- Explain employee rights relating to holiday and working hours.
- Describe organizational procedures for equality and diversity, including relevant documentation.

Key terms

Remuneration Employee's compensation for work done for the employer. This is usually a wage. It can also include complementary benefits like free health insurance.

Contract of employment An agreement between an employer and an employee.

Terms of contract The employee's employment rights, responsibilities and duties set out in the contract of employment.

Discrimination Prejudicial treatment of an individual based on their race, gender or religion.

Trade union An organization made up of members of workers. One of a trade union's main aims is to protect and advance the interests of its members in the workplace.

EMPLOYMENT RIGHTS

This chapter enables you to gain an understanding of the role of an employee in an organization trading in the retail motor industry. It covers many of the aspects required by an apprentice within the automotive industry, which form part of an apprenticeship framework.

In this section you will be exploring certain rights you are entitled to when entering employment which ensures fair treatment for all.

 **TIP** Employment laws are subject to change and the latest information and facts should be checked with the appropriate organization.

Pay (wages or remuneration)

In April 1999 a structure for the payment of a national minimum wage was introduced to enable low paid workers a guaranteed minimum amount of pay. This is age dependent and is subject to change when necessary. Employers are required by law to pay their employees no less than the minimum wage.

 **www** http://www.direct.gov.uk

 **Minimum wages**

Use the web link above to find out what the minimum wage per hour is for a:

a   16–17 year-old worker _____

b   18–20 year-old worker _____

c   21 and over worker _____

d   Apprentices under 19 or in their first year of training _____

Contract of employment

A **contract of employment** is an agreement between the employer and the employee. There are two forms of employment agreement:

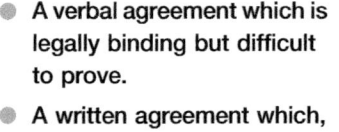

- **A verbal agreement which is legally binding but difficult to prove.**
- **A written agreement which, when signed by both parties, forms a provable, legally binding record.**

The contract initially does not have to be in writing, although within 2 months of starting employment a written statement of terms must be issued.

Contract of employment – Terms of the contract

What should the contract contain? List the main points:

_____

_____

_____

_____

_____

_____

_____

_____

_____

_____

_____

_____

When working for an employer, how long would it be before an employer must provide you with a written contract of employment stating terms of employment?

_____

Holidays

Read the following information and complete the activities.

When you are in employment you are entitled to a minimum of 5.6 weeks statutory holiday a year.

This needs to be worked out by the number of days worked a week.

Complete the number of days of holiday entitlement below:

- Working a 5 day week = _____
- Working a 2.5 day week = _____
- Working a 3 day week = _____

The maximum amount of statutory paid holiday you can be entitled to is 28 days, even if you work more than 5 days a week.

A contract of employment issued by the employer may grant you more than the minimum amount of paid holiday, for example generous employers may give 35 days paid holiday as an incentive to attract workers to their employment.

The contract of employment cannot give less than the minimum, e.g. if your contract shows 15 days paid holiday and you work 5 days a week then you are entitled to 28 days paid holiday, regardless.

Bank and public holidays

What is the difference between public holidays and bank holidays? _____

_____

There is no automatic right to take bank or public holidays off work with or without being paid. Your contract of employment should make this clear.

Unless your contract of employment specifies that you will be paid public and bank holidays on top of your normal holidays, the employer can do any of the following:

- **Ask you to work bank or public holidays.**
- **Allow you to take bank holidays off but not pay you. This will not affect your holiday allowance at other times.**
- **Allow you to take bank holidays off and pay you, but count this as part of your annual holiday entitlement.**

Why might employers offer more than the statutory amount of holiday entitlement?

_____

Circle the correct statement about a worker's rights to take public and bank holidays off:

a   Must be paid by the employer.

b   The employer cannot ask you to work.

c   If taken as holiday can be taken out of the annual holiday entitlement.

Working time

What is the maximum number of hours per week that an adult can be forced to work?

_____

Are lunch and other breaks included in the maximum weekly hours an adult works? _____

 **www** **http://www.direct.gov.uk** – search working time limits

How long can young workers (16–19 year-olds) be made to work in a week?

_____

How long do young workers have to work to be entitled to a break?

_____

What is the minimum length of break?

_____

 **Rest breaks**

Complete the blanks using information from **http://www.direct.gov.uk**.

As an adult worker (over 18), you will normally have the right to a _____ minute rest break if you are expected to work more than _____ hours at a stretch.

Daily rest – a break between working days

If you are an adult worker you have the right to a break of at least _____ hours between working days. This means as an adult worker, if you finish work at 8.00 pm on Monday you should not start work until _____ on Tuesday.

Equality Act 2010

This Act has brought together legislation already in place regarding equal opportunities.

This is a new piece of legislation (law) which makes it unlawful (against the law) to discriminate against any of the nine 'protected characteristics'.

 **www** **http://www.equalityhumanrights.com**

http://www.tuc.org.uk

List the nine protected characteristics:

1 _____

2 _____

3 _____

4 _____

5 _____

6 _____

7 _____

8 _____

9 _____

The characteristics above are used to ensure that everybody is treated fairly and must not be discriminated against.

There are different types of discrimination covered by the Equality Act 2010:

- **Direct discrimination**
- **Indirect discrimination**
- **Discrimination arising from disability**
- **Harassment**
- **Victimization**
- **Failure to make reasonable adjustments in order to accommodate a person's disability**

Name the two other types of discrimination described below:

_____ discrimination: someone who wrongly perceives them to have one of the protected

characteristics.

_____ discrimination: When someone connected to the discriminated person is also

affected. This could include the parent of a disabled child or adult or someone else who is caring

for a disabled person.

Give an example of positive discrimination relating to sex or gender in an automotive environment.

_____

Trade unions

A great deal of ignorance and misunderstanding around the rights and responsibilities of trade unions still remain.

- **Everyone has a right to join – or not to join – a** trade union **of their choice and not to be discriminated against as a consequence of that decision.**

 http://www.tuc.org.uk/tuc/unions_main.cfm

Using the website opposite list FIVE main trade unions for the motor and transport industries:

1 _____

2 _____

3 _____

4 _____

5 _____

Regarding employment and unions, what does a 'closed shop' refer to?

_____

How does the law regard this? _____

 Use this website to research the main roles of trade unions.

http://www.direct.gov.uk

Suggest three benefits of being a member of a trade union.

1 _____

2 _____

3 _____

 **http://www.tuc.org.uk/**

http://www.tgwu.org.uk/

www.acas.org.uk

How do you join a trade union?

_____

Multiple choice questions

Choose the correct answer from a), b) or c) and place a tick [✓] after your answer.

1 **What form can a contract of employment take?**

a) Verbal [ ]

b) Written [ ]

c) Verbal or written [ ]

2 **When should a written statement of employment terms be issued?**

a) After 3 weeks of employment [ ]

b) After 2 months of employment [ ]

c) After 6 months of employment [ ]

3 **At which age is the maximum amount of minimum wage payable?**

a) 21 and over [ ]

b) Over 18 [ ]

c) 20 years old [ ]

4 **Which of the following websites are best to use for employment issues?**

a) **www.hse.gov.uk [ ]**

b) **www.direct.gov.uk [ ]**

c) **www.motor.org.uk [ ]**

5 **How many protected characteristics are stated in the 2010 Equality Act?**

a) 8 [ ]

b) 9 [ ]

c) 10 [ ]

SECTION 3

Health and safety practices in vehicle maintenance

USE THIS SPACE FOR LEARNER NOTES

_____

_____

_____

Learning objectives

After studying this section you should be able to:

- Understand the requirements of the law when working in an automotive environment.
- Understand safe procedures for a range of workshop activities.
- Identify types of safety signs and their meanings.
- Complete safe lifting techniques for various types of load.
- Identify correct handling procedures for hazardous substances in an automotive workshop.
- Understand fire prevention and evacuation procedures.

Key terms and abbreviation buster

PPE   Personal protective equipment.
COSHH   Control of Substances Hazardous to Health.
HSE   Health and Safety Executive.
HASAWA   Health and Safety at Work Act (1974).
Hazard   Potential to cause harm or injury.
Risk   Likelihood or chance of harm being caused.

Health and safety is essential in all areas of work and the automotive industry is known as a high risk business because of the type of work involved.

With suitable training and supervision the likelihood of injury or damage can be substantially reduced. This chapter will focus on the main areas of safety concerning anyone starting in the industry.

C	G	N	T	S	A	E	W	P	A	O	N
T	S	O	L	D	T	E	S	A	P	V	S
E	X	T	I	N	G	U	I	S	H	E	R
S	G	I	O	A	X	A	I	E	L	R	U
I	T	C	A	T	H	O	A	L	O	A	T
A	O	E	O	S	L	H	S	G	E	L	E
P	S	S	O	E	E	A	S	G	L	L	S
R	E	G	U	L	A	T	I	O	N	S	A
N	M	F	O	X	T	E	T	G	C	X	E
A	W	A	S	A	H	F	A	V	R	E	A
R	N	P	O	E	G	T	H	E	G	N	U
S	A	O	L	F	L	S	I	S	H	P	I
H	A	Z	A	R	D	X	R	T	F	A	

COSHH
HASAWA
PPE
GOGGLES
AXLESTANDS
OVERALLS
FOAM
FUEL
NOTICES
REGULATION
EXTINGUISHER
HAZARD

HEALTH AND SAFETY EXECUTIVE (HSE)

The HSE is the national independent watchdog for work-related health, safety and illness. The HSE inspectors visit an organization for any of the following reasons:

- **To investigate companies where health and safety has been reported as poor.**
- **To ensure correct observance of health and safety standards.**
- **To investigate accidents in workplaces.**

The HSE has an extensive website giving support and advice to both employers and employees. There is a dedicated section for health and safety in the motor vehicle repair (MVR) industry, which focuses on issues in the garage trade.

www HSE website

http://www.hse.gov.uk

Visit this website for information to complete activities relating to health and safety.

LEGISLATION – LAWS SOMETIMES CALLED ACTS

Health and Safety at Work Act 1974 (HASAWA)

This act was introduced in 1974 to ensure all employees were covered by health and safety legislation in all places of work. HASAWA lays down responsibilities for employers and employees.

Since then, numerous other acts and regulations have come into force. Details can be found on notices at work and on the HSE website. A *small* selection of acts related to your sector are shown below:

Electricity at Work Regulations 1989

Example – mains powered equipment, e.g. angle grinder

Personal Protective Equipment at Work Regulations 1992

Refers to types of protection and their care and use

Provision and Use of Work Equipment Regulations 1998 (PUWER)

Examples – bench grinders, drills, hydraulic presses

Using the HSE website, write down the acts or regulations relating to:

● The use of vehicle ramps, engine hoists, slings and lifting chains. _____
_____

● Substances that may affect our health, e.g. battery acid or exhaust fumes _____
_____

● Safe lifting using our body _____

In your workshop or workplace (a college or training provider is known as a workplace under health and safety terms), locate the main health and safety information and notices on display.

Sketch a plan of your workshop and indicate the location of the main health and safety notices, fire exits, fire extinguishers, fire alarm points and fixed items of equipment.

Workshop plan

In the table below, list four health and safety responsibilities that you *must do* and give a practical example of when each responsibility would apply.

Health and safety responsibility	Example of when the responsibility would apply
● _____ _____ _____	● _____ _____ _____
● _____ _____	_____ _____
● _____ _____	_____ _____
● _____ _____	● _____ _____
_____	_____

HAZARD SPOTTING

What is a hazard? _____

Circle all the hazards you can find on the picture below.

Discuss with your class what you found.

Visit your workshop and list four possible hazards that may be commonly found. Give an example of how each hazard has been made safer.

1 Hazard _____

Made safer by _____

2 Hazard _____

Made safer by _____

3 Hazard _____

Made safer by _____

4 Hazard _____

Made safer by _____

ACCIDENTS

An accident is an unplanned event; we do not get up in the morning and plan to have an accident.

Causes of accidents in a garage or vehicle workshop often fall into three categories:

- **Human error**
- **Equipment failure**
- **Unsafe working conditions and practices**

State TWO causes in each category.
One cause is completed for you.

Human error:

1 Daydreaming.

2 _____

3 _____

Equipment failure:

1 Lack of maintenance of lifting and electrical equipment.

2 ______________________________

3 ______________________________

Unsafe working conditions:

1 Poor lighting.

2 ______________________________

3 ______________________________

PERSONAL PROTECTIVE EQUIPMENT (PPE)

Use the word scramble to identify types of PPE found in automotive workshops.

Use the images to help you. Give an example of when each type of PPE would need to be used.

GOSEAGELLGRC □□□□□ □□□□□□□

BRGLEOUSREBV □□□□□□ □□□□□□

VLLOARES □□□□□□□□

AREDDREEENFS □□□ □□□□□□□□□

AGSELKIDNWM □□□□□□□ □□□□

OYTAEOBFSST □□□□□□ □□□□□

RTAVSEEELOGHL □□□□□□□ □□□□□□

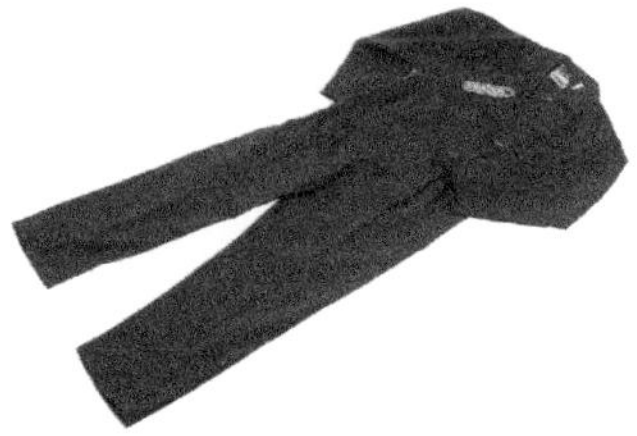

____

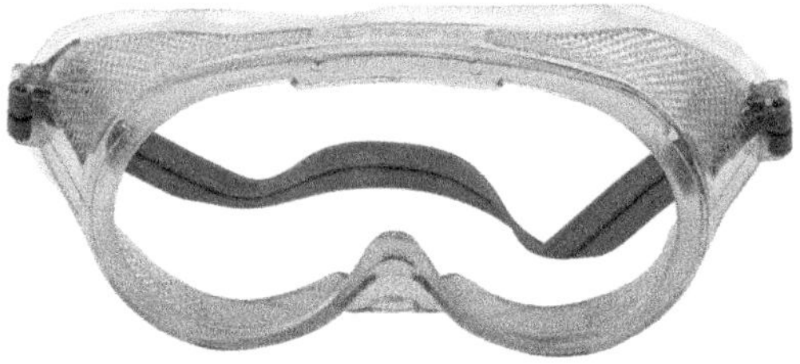

____

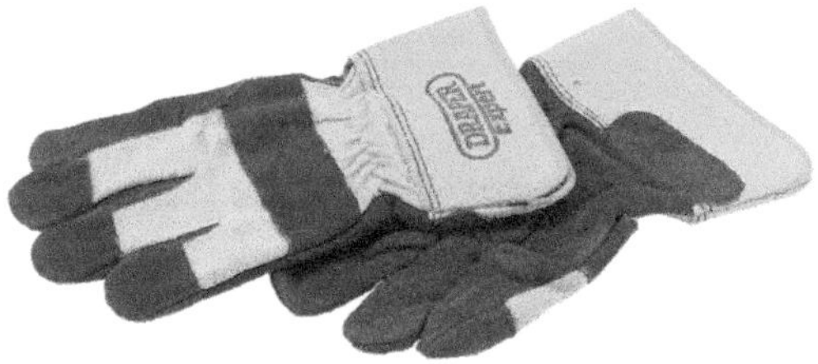

____

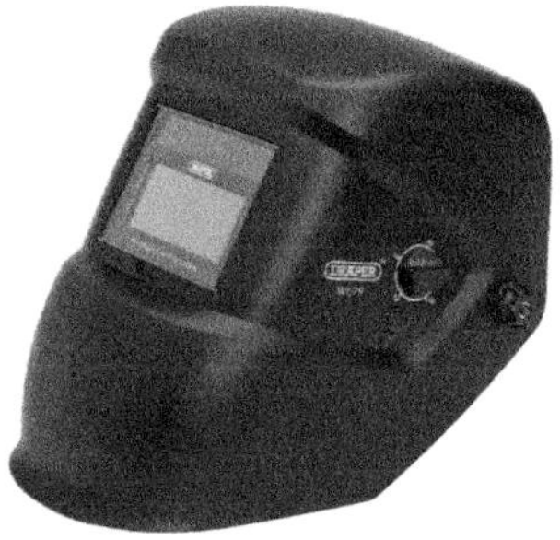

____

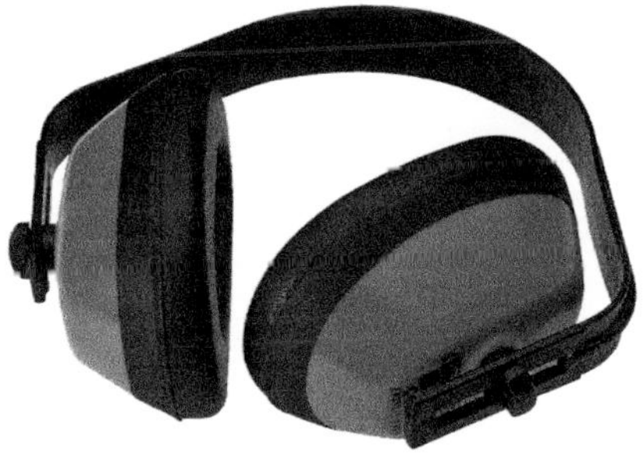

____

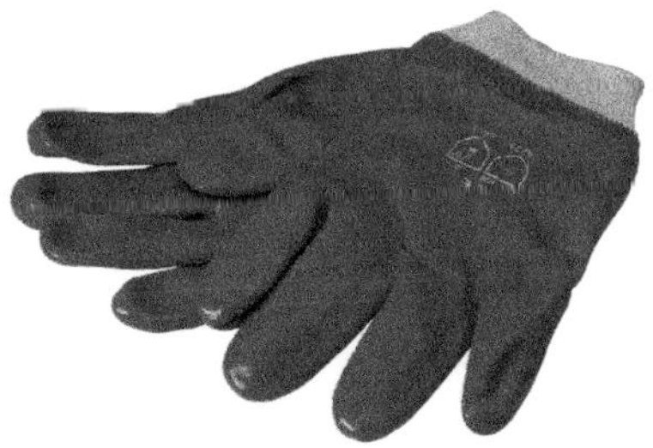

____

All above images supplied by Draper Tools Ltd

_____

Consider the typical workshop tasks shown in the following table:

1 State the PPE common to them all.

_____

2 Alongside each task state any special PPE required.

Task	PPE required
Draining engine oil	
Filling battery acid	
Using a pillar drill	
Removing a section of exhaust using an angle grinder	
Heating or welding	

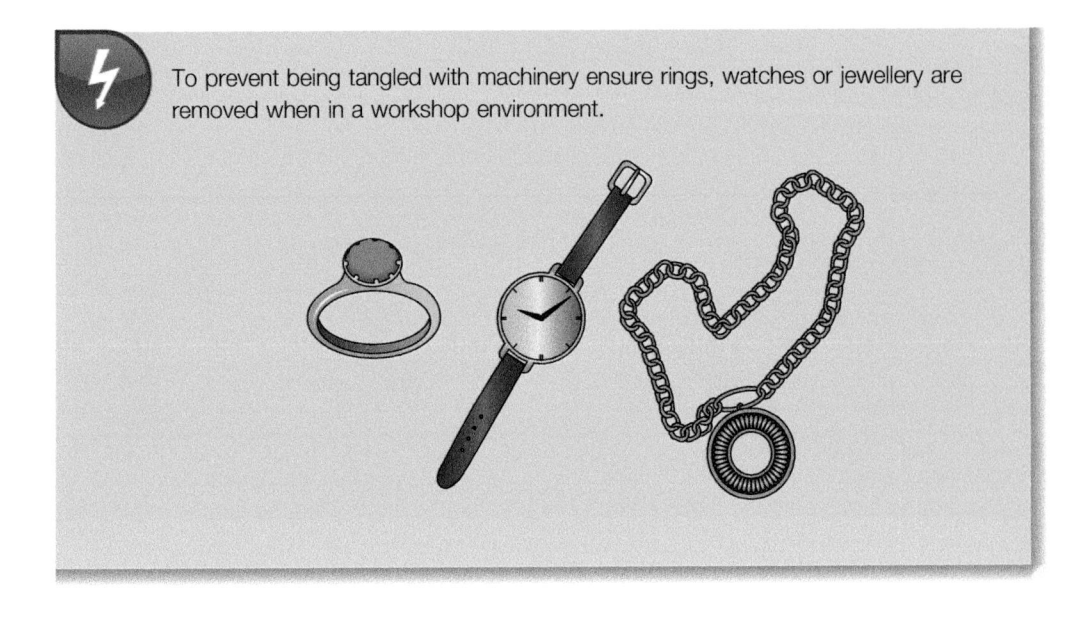

To prevent being tangled with machinery ensure rings, watches or jewellery are removed when in a workshop environment.

MANUAL HANDLING

When working in a vehicle workshop there are many situations when components or boxed parts need to be moved or lifted using manual lifting techniques.

It is important to ensure that you are trained to lift correctly to reduce the chance of injury.

When lifting is being performed overalls should be worn to prevent any loose clothing being caught.

State which other item of PPE would be required. _____

Follow the method of a squat lift in the drawing below and complete the procedure using these words:

gloves   **close**   **back**   **legs**
head   **feet**   **apart**   **knees**

1 Stand as _____ to the load as possible, with _____ shoulder width

_____ if possible.

2 Bend your _____ keeping your _____ straight.

3 Grip the load fully, using _____ if there are sharp edges.

4 Raise your _____.

5 Lift by straightening your _____. Keep the action smooth.

6 Hold the load _____ to your body.

TIP If the load is awkward ask someone to assist the lift. With heavy loads, use lifting equipment.

LIFTING EQUIPMENT

There are many tasks in a vehicle workshop where lifting equipment is essential, such as:

- **Raising vehicles to gain access to components**
- **Removing power units**
- **Removing transmission components.**

Trolley jack

When raising vehicles off the ground in a workshop, a trolley jack is normally used for speed and safety.

Before using the trolley jack you are required to carry out safety checks.

In the areas circled state the checks required on this hydraulic trolley jack.

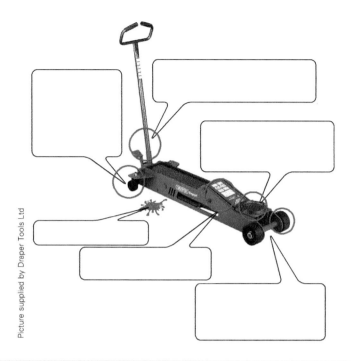

Picture supplied by Draper Tools Ltd

⚡ If the jack is faulty, do not use it and report the fault to your workshop supervisor.

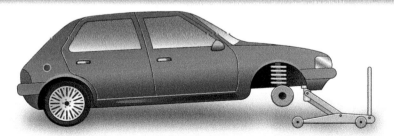

What's missing here?

The diagram on page 21 shows a vehicle jacked up ready for inspection. Show on the diagram the correct positions of the two necessary items required to ensure safe working under the vehicle.

Complete the list of important features relating to axle stands by selecting the missing words below:

Picture supplied by Draper Tools Ltd

Axle stands

SWL	height	material
feet	centrally	welds

- Both axle stands must be the same type and set to the

  same _____ when supporting the front or rear of the

  vehicle.

- The load must be within the _____.

- Height adjustment pins must be the correct _____ and fitted _____ in the tube.

- _____ must be checked before use for signs of cracking.

- All _____ must be touching the floor.

Vehicle hoists (commonly known as ramps)

To allow the underside and suspension components to be easily removed and inspected the vehicle hoist is used by most automotive businesses.

Two main types of hoist are popular.

4-Post ramp

2-Post or wheel free ramp

In your college or workplace ask to be trained in the use of these ramps and list the FIVE essential checks.

Ramp type: 4-post

Make:

SWL:

Five checks:

1 _____

2 _____

3 _____

4 _____

5 _____

Ramp type: 2-post

Make:

SWL:

Five checks:

1 _____

2 _____

3 _____

4 _____

5 _____

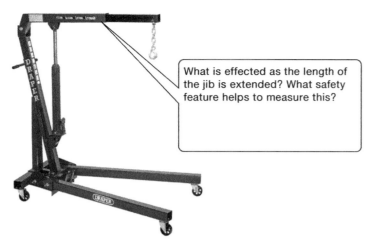

What is effected as the length of the jib is extended? What safety feature helps to measure this?

Engine Hoist

Lifting chains

 Check links for wear or cracks.

 With the help of your supervisor, try using this equipment to lift an engine and transport it to another location in the workshop.

 Keep the engine close to the ground when moving around the workshop.

SAFE USE OF WORK EQUIPMENT

Electrical safety

What are the two main dangers caused by electricity?

● _____

● _____

Always carry out basic visual checks before using mains power tools.

Each appliance should be checked for electrical safety and have a portable appliance test (PAT) carried out regularly. A label on the equipment will indicate that it has been checked for electrical safety.

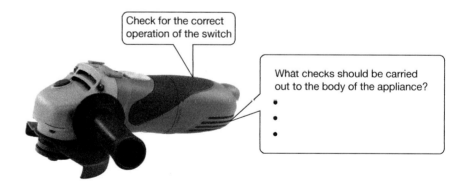

Check for the correct operation of the switch

What checks should be carried out to the body of the appliance?
●
●
●

All above images supplied by Draper Tools Ltd

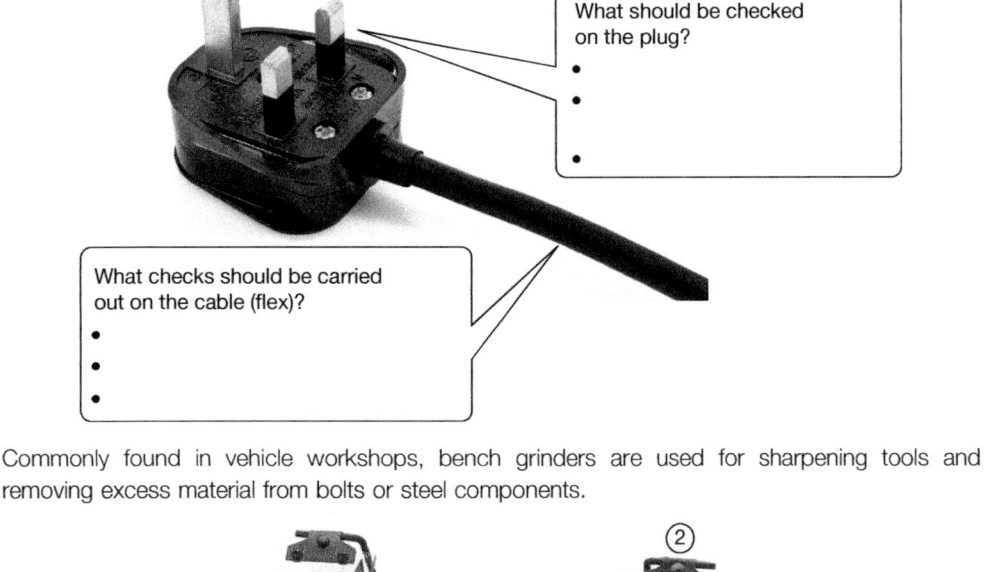

What should be checked on the plug?
•
•
•

What checks should be carried out on the cable (flex)?
•
•
•

Why must this belt guard be in place when the drill is operating?

What is the purpose of this chuck guard?

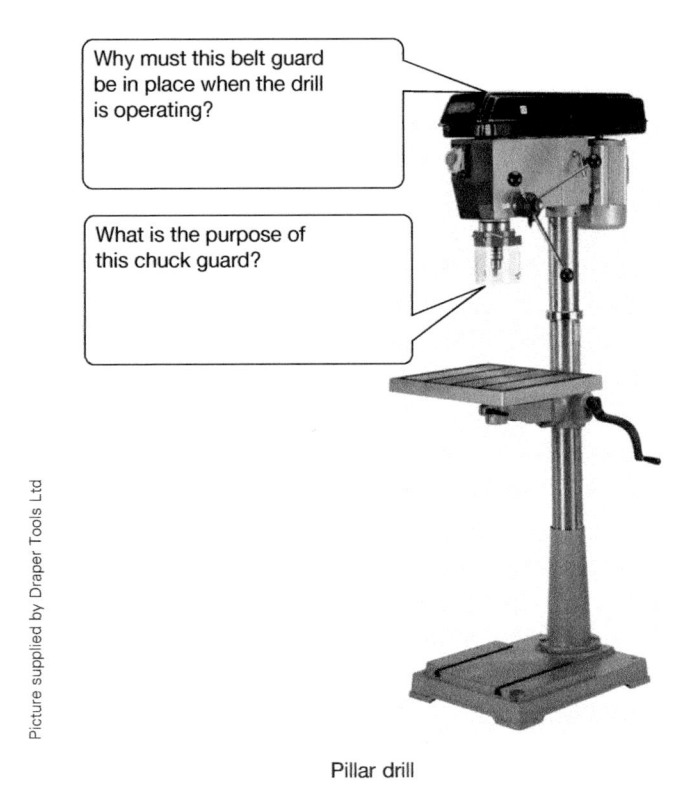

Picture supplied by Draper Tools Ltd

Pillar drill

Commonly found in vehicle workshops, bench grinders are used for sharpening tools and removing excess material from bolts or steel components.

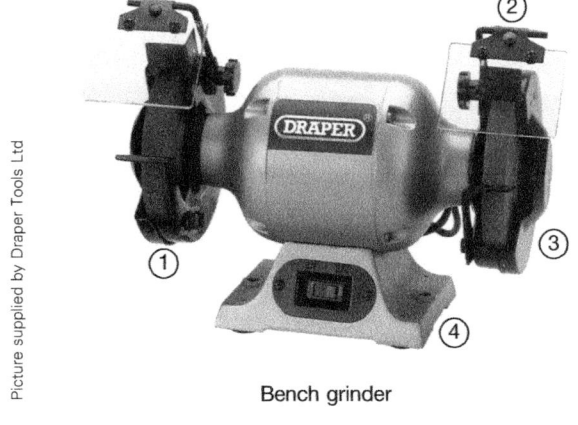

Picture supplied by Draper Tools Ltd

Bench grinder

Complete the statements about safety which relate to the numbered parts of the bench grinder above.

| times | rest | bolted | close | guard | shield |

1 The tool _____ must be adjusted as _____ to the grinding wheel as possible.

2 If an eye _____ is fitted it should be in position when the grinder is in use.

3 The _____ must be in position at all _____, in case the wheel bursts.

4 The grinder must be _____ down to a bench or solid stand.

HANDLING HARMFUL SUBSTANCES

In a vehicle workshop many substances are present which can affect our health. There are two forms:

● Manufactured substances that are packaged and labelled.

State THREE examples – _____

● Substances that may be present but not packaged.

State THREE examples – _____

The regulation concerning harmful substances is called:

Control **of S**ubstances **H**azardous to **H**ealth commonly known as COSHH.

The COSHH regulation places responsibility on employers and suppliers to reduce the **risk** of injury or ill-health caused by any substances used.

Warning symbols, such as those shown below, are used on packaging.

State the meaning of these symbols:

    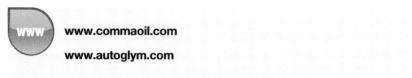

HSE

_____   _____   _____   _____   _____

_____   _____   _____   _____   _____

Locate FIVE products in your garage or workshop that are commonly used and from information on their labels complete the table below. The first line is completed for you.

Products	Warning(s)	Safety precautions
Brake fluid	Irritant	Wear nitrile gloves

The symbol used to indicate irritant and harmful substances is the same, find the symbol and draw it in the box.

☐

Any substance supplied must have a manufacturer's safety data sheet (MSDS) giving more detailed information about the product.

To see samples of a MSDS, visit a product manufacturer's website of your choice or view the suggested sites below:

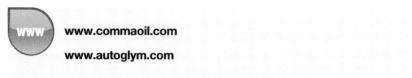

WWW **www.commaoil.com**

www.autoglym.com

FIRE PREVENTION AND EMERGENCY PROCEDURES

When you start at college or a new place of work, you will be told what to do in the event of a fire. Your college or employer will have taken precautions to prevent a fire occurring and will have a planned procedure to be carried out if a fire starts.

Write the THREE elements needed to start a fire on the triangle below and explain what you would do to each element to stop a fire.

In groups, discuss situations in a garage that could lead to a fire starting.

List THREE operations that could be particularly hazardous:

1 _____

2 _____

3 _____

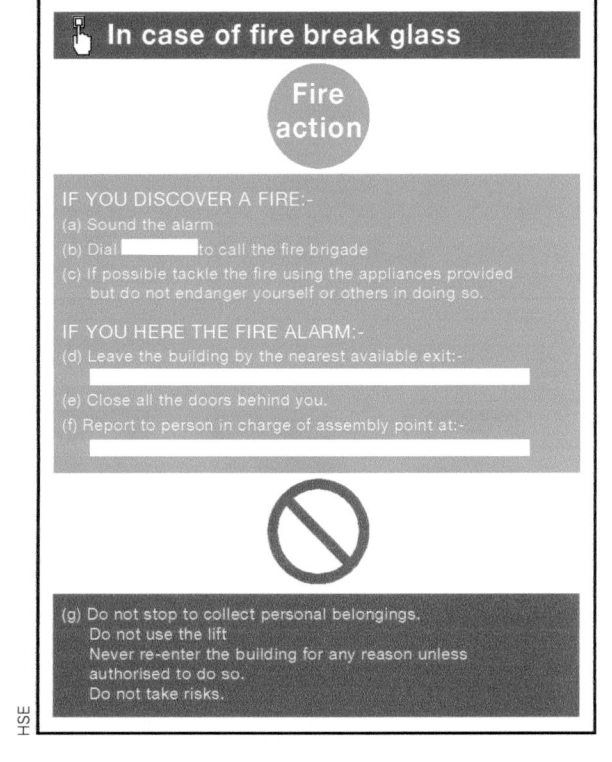

www **www.firemart.co.uk**

What to do in the event of a fire

In your college or workplace there will be information on what do in the event of fire.

In case of fire break glass

Fire action

IF YOU DISCOVER A FIRE:-
(a) Sound the alarm
(b) Dial ▮▮▮▮ to call the fire brigade
(c) If possible tackle the fire using the appliances provided but do not endanger yourself or others in doing so.

IF YOU HERE THE FIRE ALARM:-
(d) Leave the building by the nearest available exit:-
_____
(e) Close all the doors behind you.
(f) Report to person in charge of assembly point at:-
_____

(g) Do not stop to collect personal belongings.
Do not use the lift
Never re-enter the building for any reason unless authorised to do so.
Do not take risks.

HSE

How should you notify others if you discover a fire?

_____   _____   _____

Where are the fire exits?

Mark workshop fire exits and any alarm points on your workshop diagram on page 17.

Draw the sign used to identify a fire exit in the space below.

HSE

Find your fire assembly notice and ask your tutor or supervisor where to assemble if the alarm is raised.

List THREE locations in your college or workshop area and state their relative fire assembly points in the table below.

Location	Fire assembly area

Fire extinguishers

Use local fire extinguishers or search for information to complete the table on page 27, indicating uses of each type. Tick the types of fire each extinguisher can be used on.

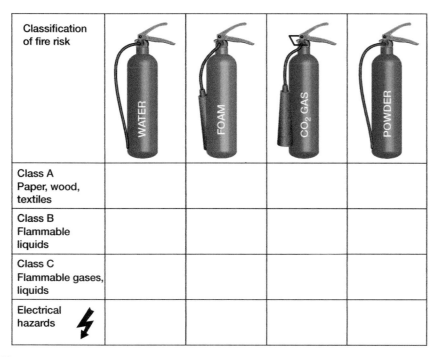

Classification of fire risk	WATER	FOAM	CO₂ GAS	POWDER
Class A Paper, wood, textiles				
Class B Flammable liquids				
Class C Flammable gases, liquids				
Electrical hazards				

Caution

Why can water or foam not be used on electrical fires?

_____

What would happen if water was used on a fire involving a flammable liquid?

_____

Safety signs

Safety signs are used as a quick method to communicate important messages.

Four colours are used to help identify the meaning of the safety sign quickly.

Blue = Mandatory – means 'must do'
Red = Prohibition – means 'do not'
Yellow = Warning – means 'be careful'

Green = Safe condition – Means 'ok to do or go to'

Find the following signs around your workshop and work area and colour in or write in the correct colours. State what each sign are telling you.

HSE

_____ _____ _____ _____

HSE

_____ _____ _____ _____

WORKING ON VEHICLE ELECTRICAL SYSTEMS

When removing components such as starter motors and alternators, or carrying out major engine or electrical work, it is necessary to disconnect the battery.

It is good practice to remove the battery when carrying out major work.

State the correct procedure for battery removal and replacement using the following words in the correct place:

positive   **negative**

Removal procedure

- Ensure ignition and all circuits are switched off
- Disconnect the earth terminal – _____
- Disconnect the _____ terminal
- Release the battery clamp

Replacement procedure

- Clamp battery in place
- Connect the _____ lead
- Connect the _____ lead

 When dealing with vehicle electrical systems, remove metal strapped wrist watches, metal bracelets and restrain any necklaces. No jewellery should be worn in the workshop at any time.

TIP Classic cars may have a positive earth, ask if you are unsure.

GOOD HOUSEKEEPING

A garage workshop needs to be kept clean and tools and equipment stored correctly to ensure a safe and efficient working environment is maintained.

Describe how this hazard should be dealt with.

_____

_____

_____

_____

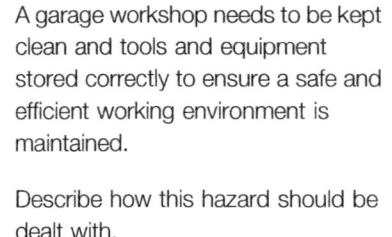

Oil spilled on the floor is a slip hazard

State how you would improve each housekeeping situation described below.

Worn brake discs left on the floor or on benches.

Tools lying on the floor.

Engine coolant spilt on the floor.

Air lines trailing on the floor.

WASTE DISPOSAL

Regulations are in place to prevent contamination occurring from harmful waste and to protect the environment.

Hazardous waste must be stored and disposed of correctly and taken away by a: _____.

Used engine oil and oil filters are classed as hazardous waste.

Who takes away oil and filters from your college/training provider/workplace?

_____

Using the Internet, search for a company specializing in garage waste disposal and find FOUR more examples of harmful waste:

1 _____

2 _____

3 _____

4 _____

Multiple choice questions

Choose the correct answer from a), b) or c) and place a tick [✓] after your answer.

1 **Which of the following responsibilities regarding health and safety is expected of an employee?**

   a) Provide your own personal protective equipment [ ]

   b) Repair any faulty equipment before using it [ ]

   c) Follow the training that the employer has given you [ ]

2 **Which of the following is a human cause of accidents?**

   a) Working in poor lighting conditions [ ]

   b) Fooling around at work [ ]

   c) Using unguarded machinery [ ]

3 **This diagram shows a one person lift. What is number 4 indicating?**

   a) Keep your head up [ ]

   b) Look at your load [ ]

   c) Look where you are going [ ]

4 **Explain the meaning of this safety sign.**

   a) No smoking [ ]

   b) No matches allowed [ ]

   c) No naked flames [ ]

5 **What three elements are required for a fire to start?**

   a) Air, heat and ignition [ ]

   b) Fuel, heat and oxygen [ ]

   c) Spark, fuel and heat [ ]

6 **What type of fire extinguisher has a black band marking?**

   a) Water [ ]

   b) Dry powder [ ]

   c) Carbon dioxide [ ]

7 **Which of the following must be used to support a vehicle when working underneath?**

   a) Trolley jack [ ]

   b) Axle stands [ ]

   c) Ramp [ ]

8 **Which of the following tasks would require you to wear safety goggles?**

   a) Draining engine oil [ ]

   b) Using a bench grinder [ ]

   c) Replacing an air filter [ ]

9 **When working with workshop chemicals and products, what does this symbol indicate?**

   a) Toxic [ ]

   b) Harmful or irritant [ ]

   c) Explosive [ ]

10 **Which of the following safety checks would need to be carried out on a 240 volt mains angle grinder?**

   a) Remove the plug top and check the fuse [ ]

   b) Check the lead for signs of cuts or splits [ ]

   c) Check the grinding disc for wear [ ]

SECTION 4

Tools, equipment and materials

USE THIS SPACE FOR LEARNER NOTES

_____

_____

_____

_____

_____

_____

_____

Learning objectives

After studying this section you should be able to:

● Identify a variety of tools and equipment used in a motor vehicle work.
● Be aware of how these tools and equipment are used.
● State how to care for and maintain tools and equipment.
● Identify the types of materials used in the construction of vehicles.
● Identify methods of fastening and joining materials and components.
● Identify materials used in the routine maintenance of vehicles.

Key terms

Ferrous A metal which contains iron, making it magnetic.
SWL Safe Working Load. The maximum weight that can be lifted or supported.

www www.draper.co.uk

www.britool.com

www.snapon.com

www.gedoreuk.com

```
R G N I T T E S O M R E H T D
L E S I H C U D I N N I A E N
N T T L E O I I E E M F C H A
M R D E R E P M E T I E K C L
D T L R M C P H I L I P S T U
V I E I O O V I E C U V A A T
H F R V P R R E B B U R W R T
S T E E L D C C R P M D E E C
H I E T A L A M I N A T E D H
A E V I S E H D A M I T A N G
N C R N T S S P A N N E R I D
K I E G I S S R R T A R R R E
E V I A C A C S S P P P I G E
I P I T R L E P R P M I O E N
M E W A T G N D R T C A M G N
```

CORDLESS   GRINDER   VICE   TAP   DIE   TEMPERED

HACKSAW   RUBBER   LAMINATED   RATCHET   FERROUS

STEEL   THERMOSETTING   ADHESIVE   FILE   CHISEL

RASP SHANK TANG LAND GLASS MICROMETER VERNIER

RULE SPANNER PHILIPS THERMOPLASTIC RIVETING

TOOLS USED WHEN SERVICING VEHICLES

Anyone employed in the active repair of motor vehicles should eventually build up a comprehensive set of tools suitable for their specialized type of work. Name the tools shown:

_____   _____

State the common sizes and types of spanners used:

_____

_____

_____

_____

_____

_____

_____   _____

State the common ranges and square drives required for a socket set.

_____

_____

_____

 Locate and identify the following tools in your workshop: ¼" drive ratchet; ³/₈" drive ratchet; 6" long ³/₈" drive extension; ½" drive deep impact socket; ³/₈" drive Torx socket and a ½" drive breaker bar.

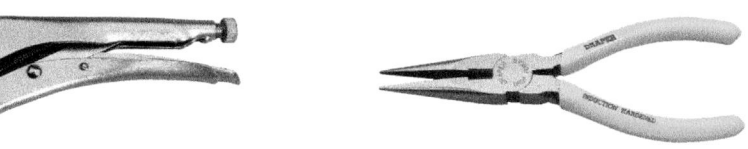

_____

Name TWO other types of pliers:

_____

_____

Screwdriver set

➕ **PHILLIPS TIP**

✳ **POZIDRIV® TIP**

✴ **TORX® TIP**

🎀 **CLUTCH TIP**

■ **SCRULOX® (SQUARE TIP)**

Various screwdriver tips

The figure opposite shows a selection of tips available. Ideally a well-equipped tool box should include a selection of these. Screwdrivers are defined by their sizes, tips and the types of fasteners they should be used with.

Sketch a screwdriver below and correctly label the tip, handle and shaft:

Complete the following table which identifies different types of screwdrivers and their use:

Screwdriver type	Use	Example of use
_____	Used to undo and tighten flat head screws.	_____
Philips	_____ _____	_____
_____	Used to undo and tighten flat cross-head screws.	_____
Torx	Used to undo and tighten torx head screws.	_____
_____	Used for speed. Different types of detachable blades can be fitted.	_____
Electrical	_____ _____	_____ _____

An impact screwdriver set is a good addition to the toolbox. These come with interchangeable heads and bits.

When would an impact screwdriver be used?

_____

Torque wrenches come in a range of square drive sizes, lengths and torque setting calibrations. What is the basic function of a torque wrench?

Torque wrench

_____

In what units are torque wrenches normally calibrated?

_____

The type of torque wrench shown above should always be unwound after use. This is so that the internal spring does not weaken. What could a weak internal spring cause? _____

_____

Select a number of different size torque wrenches. Look at the different scales on the wrench. Discuss, either as a group or in pairs, what these mean (e.g. Nm etc).

Choose three different vehicles in the workshop. Look up the technical data (from a manual or electronic database) for these vehicles and determine the torque settings for four different components (including wheel nuts/bolts). Make sure they are tightened to their correct setting.

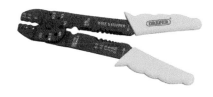

Name this tool and state its various functions:

_____

Name this tool and state TWO typical uses: _____

_____

Hammers

The hammer is a very common tool in all trades. In motor vehicle work, different types are used to suit different purposes.

Name the types of hammers shown and state a typical use for each different face:

Picture provided by permission of Snap-on Industrial

Picture provided by permission of Snap-on Industrial

_____

_____

Picture provided by permission of Snap-on Industrial

_____

_____

_____

_____

All above images supplied by Draper Tools Ltd unless stated

Saws

The hacksaw is the most common type of saw used for cutting metals. The frame may be as shown in the figures below and may be adjustable or non-adjustable.

Adjustable large hacksaw   Non-adjustable junior hacksaw

Hacksaw blades are made from heat treated high carbon steel and are classified according to their length and number of teeth per 25 mm (known as the pitch). The pitch can range from 18 to 32. For general purpose use, the pitch is normally _____.

The choice of blade depends upon the shape and type of material being cut.

Blades with teeth of 'fine pitch' are used for cutting _____

Which types of materials are saw blades with teeth of a 'coarse pitch' used for?

_____

Insert the missing words into the following paragraph using words from the word bank below (note: there are two extra distracter words):

offset	clearance	wedge	shearing	teeth
handle	cut	set	blade	in-line

The individual teeth on a hacksaw _____ form a pointed '_____' which digs into the metal and produces a _____ action. The teeth on the blade are _____ from one another. This offsetting is called the _____ of the _____. This is so the blade has _____ to move freely through the _____ groove.

Which way round should the blade be fitted into a hacksaw frame? _____

_____

On which stroke does the cutting action occur? _____

Use a ball pein hammer to make a gasket. For example: the gasket could be made for an inlet manifold, thermostat housing or a half shaft flange.

Describe below how to make a gasket using a hammer:

_____

_____

_____

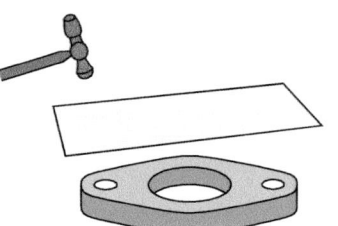

Using a ball pein hammer to make a gasket

Punches

Name the types of punches shown below and where they would be used:

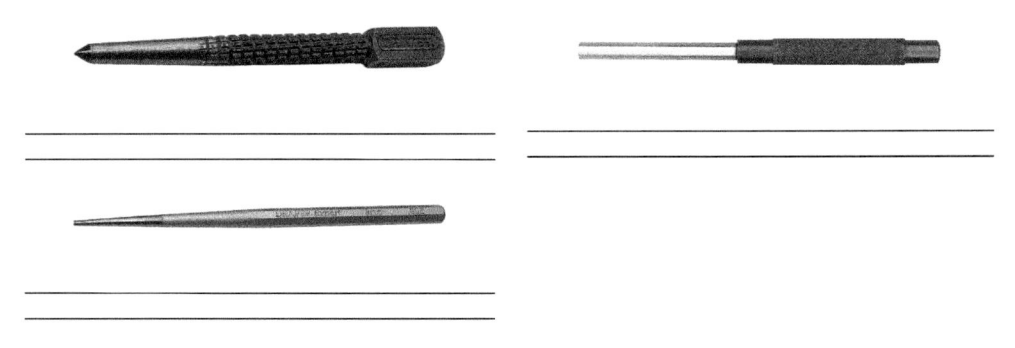

_____   _____

_____   _____

_____

_____

Give THREE causes of saw blade breakage:

1 _____

2 _____

3 _____

 TIP When cutting thin sheets it is good practice to have at least three teeth in contact with the metal.

Shears for cutting metal

Tin snips are commonly used for cutting thin sheet steel. The snips may be flat or curved-nosed. They are used very much like a pair of scissors.

For cutting large sheets of metal (or metal too thick for snips) a guillotine or bench shears may be used.

Observe a demonstration cut with tin snips and say what happens to the narrow waste side of the metal. _____
_____

Straight tin snips

Chisels

The cutting action of a chisel works on the principle of forcing a wedge into the material to shear off any unwanted material.

Chisel

Hammer blows cause the pointed 'wedge' to 'shear' through the metal. The depth of cut is maintained by holding the chisel at the correct angle (angle of inclination). Three other important factors that affect the efficiency of the cutting action are shown opposite.

All above images supplied by Draper Tools Ltd

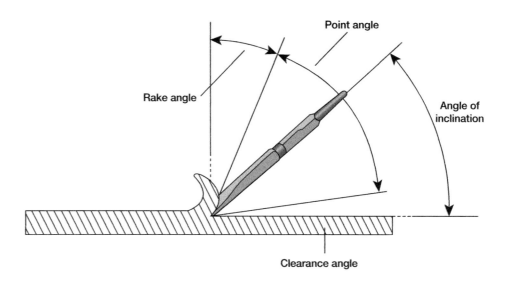

Point angle

Rake angle

Angle of inclination

Clearance angle

 TIP Always use a sharp chisel cut to the correct angle. It makes the task easier and it will help to produce a better finish.

Chisels with point angles in the region of 55–65° are used for cutting relatively hard materials and the tool point is quite strong. For cutting softer materials the point angle is reduced.

When chiseling low carbon steel the point angle needs to be ___ and when chiseling aluminium, the point angle needs to be ___,

 Wear goggles to protect the eyes when using a chisel.

The common engineering chisel is often called a _____. They are usually made from high carbon steel which is hardened only at the pointed end. Why is the head left soft?

_____

 Never allow a chisel to become excessively 'mushroomed' as fragments of the metal could fly off when being hit with a hammer. Remove the excess 'mushroom' on a bench grinder (always wear goggles) before use.

Right Wrong

When chiselling and striking with the hammer, on what part of the work should the eyes be focused?

_____

Files

The cutting action of a file is similar to that of a chisel or hacksaw. Each tooth on a file is a tiny cutting blade. Files are classified according to length, shape, grade of cut and type of cut.

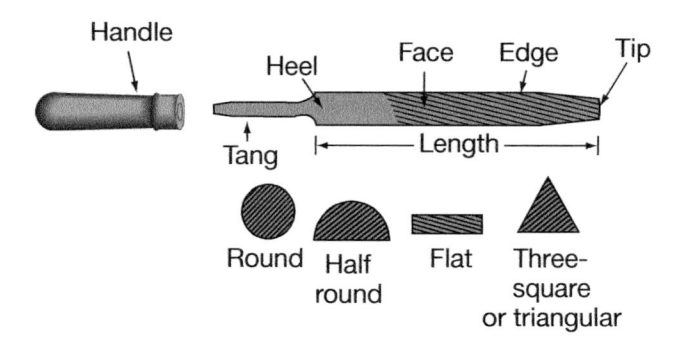

Files common in a variety of shapes

As can be seen above, there are file shapes to suit any work situation.

The file teeth or angles of file teeth opposite. Correctly label the types of file teeth shown from the following:

Single cut   **Rasp**   **Double cut**

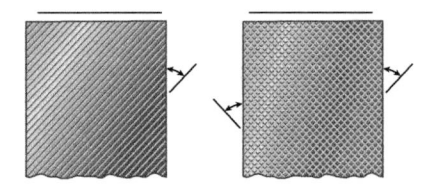

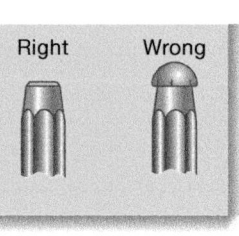 Never use a file without a handle. The tang is sharp and can be dangerous. Never use a file as a lever. They can easily break and a piece may fly into someone's eye.

Files are graded according to the number of teeth per centimetre, that is, their roughness or smoothness. Four grades of file between the rasp and dead smooth are shown below. Draw lines to the statement which correctly describes their use:

Rough	Rough cutting hard steel
Bastard	Finish cutting, draw filing
Second cut	Filing hard brass, rough cutting steel
Smooth	Filing soft brass, aluminium, plastics

State FOUR methods of care that help maximize the life of a file:

1 _____

2 _____

3 _____

4 _____

List FOUR other basic hand tools you consider desirable to have in a toolbox:

_____   _____   _____   _____

Drills

The most common type of drill is the twist drill. It is supplied in a number of lengths: jobber series, these are normal length drills and long series, these are stub (short) drills. Metric drill sizes range from 1 to 20 mm diameter, the smaller sizes being available in 0.1 mm steps. Name the main parts of the drills shown in the figure below, including the different types of shank:

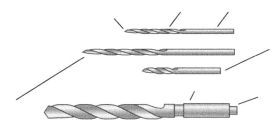

Types of drill bits

Other types of drills are shown below. Correctly label them from the following word bank:

Flat   **Counter sunk**   **Straight fluted**   **Slow spiral**   **Quick spiral**

Each of the drills shown opposite are designed for a specific use. In the following table match the correct drills to their designed use:

Drill	Designed use of drill
_____	For drilling plastics.
_____	Used for cutting sheet metal.
_____	For centring work in lathes and countersinking.
_____	For drilling copper and soft metals.
_____	Deals with hard spots in cast iron and prevents digging in soft metal.

Twist drills remove metal by using two cutting edges rotating about a centre point.

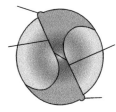 

On the figure above, indicate the cutting edges and lands. Indicate the included (point) angle. Show the direction of rotation.

The point angle necessary for general purpose work is _____.

State the purpose of the *land* on a twist drill: _____

When a drill is sharpened by grinding, it is essential to keep the cutting edges (lips) the same length and at the same angle to the drill axis. An incorrectly sharpened drill can produce an oversized hole.

TIP A quick and easy way to check the point angle of a drill is correct is to place two nuts together. The angle formed between the two nuts is virtually the correct angle required for the drill point.

As well as an incorrectly sharpened drill, there are other causes of inaccurate drilling. State THREE other causes:

1 _____

2 _____

3 _____

Insert the missing words into the following two paragraphs using words from the word bank below (note: there are two extra distracter words):

plastics	small	materials	large
oil	vertical	centre	horizontal
bits	precautions	carbon	dry
overheating	bronze	accuracy	stages

Soluble _____ (mixed with water) should be used when machine shop drilling low

_____ steel, aluminium, copper and phosphor _____. Cast iron, brass and _____

can be drilled dry. When using a hand drill on a vehicle, all _____ will be drilled _____,

so care must be taken to avoid _____.

When drilling a hole, a number of _____ need to be taken to ensure _____ of the operation:

_____ punch then drill a _____ pilot hole. Drill out the hole with larger drill _____,

in suitable _____ to the required size hole. Hold the drill _____ so that the bit is 90° to

the work face.

 **TIP** If the drill goes off centre when starting to drill a hole, draw it back to the centre with gradual pressure. If this fails to work, use a small chisel and chisel a small groove towards the centre, then re-drill.

Air tools

These are connected to an airline and are designed to make a lot of jobs quicker and easier to perform. Air tools can be used in areas where there is danger of electric shocks.

The impact wrench (see below) makes it easier to undo and tighten nuts and bolts.

Give THREE examples where an impact wrench could be used:

1 _____

2 _____

3 _____

Briefly describe why an impact wrench should not be used to tighten nuts, bolts and fasteners:

_____

_____

Impact wrench, air, ½" drive   Ratchet, air, ³/₈" drive

Instead of using hand ratchets, an air ratchet (see figure above) can be used to make the job of undoing nuts, bolts and fasteners easier and faster. The torque rating of the air ratchet needs to be checked to make sure nuts and bolts are not over-tightened.

TIP To maintain and prolong the life of air tools, they need to be regularly lubricated with the correct type of oil.

Electric powered tools

These are used for their convenience and portability.

List THREE other cordless tools likely to be used in the workshop:

1 _____

2 _____

3 _____

14.4 v Cordless hammer drill with two batteries

 It is important that you are trained how to safely and correctly use all tools and equipment. This is especially important with power tools.

Electric drills can be either hand-held or bench mounted. Both types can have variable speed settings and may have a reverse action. The bench drill is useful for accurately drilling components (the bench drill shown has 16 speeds).

Bench drill

All images in column above supplied by Draper Tools Ltd

Bench grinder

These are securely mounted to a bench. They may have an abrasive grinding wheel at each end (these may be different grades), or one end could have a wire wheel. Bench grinders rotate at very high speeds.

TIP The hardness of the metal can be affected by overheating when grinding. If it becomes too hot allow it to cool slowly, as quenching in water (cooling quickly) can make it become brittle.

As the metal is being shaped, dip it in water to prevent it from getting too hot.

State THREE uses for a bench grinder:

1 _____

2 _____

3 _____

The abrasive wheel eventually wears down and the gap between the wheel and the tool rest will increase. It is important that the tool rest is as close as possible to the grinding wheel but not touching it. The gap needs to be about 1.5 mm.

The face of the abrasive wheel must be kept square. This is done with a dressing tool, which removes some of the abrasive compound.

Do not grind on the side of the wheel as this can cause the wheel to shatter.

Check the safety shields and tool rests are fitted and secure.

Wear goggles and use the safety shield fitted to the grinder.

Angle grinders

These are portable, mains-powered grinders. The grinding disc needs to be checked before use and with the appliance unplugged from the mains. If the disc is insecure or excessively worn down, it will need to be replaced by a suitably trained member of staff.

Special-purpose tools and equipment

These are made to do a specific job on a vehicle and are numerous and varied. They range from a spark plug spanner to inner door handle removal tools.

Tick those tools below which are classed as special-purpose:

Clutch alignment tool

Philips screwdriver

3/8" drive ratchet

Piston ring compressor

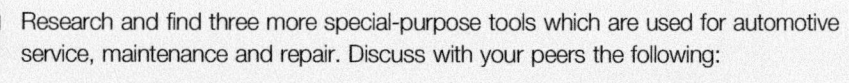

Research and find three more special-purpose tools which are used for automotive service, maintenance and repair. Discuss with your peers the following:

● What they are used for?
● How should they be cared for (e.g. calibrated, cleaned, stored)?
● How should they be used (including any safety procedures)?

Suspension coil spring compressor

5 litre measuring jug

Brake caliper piston wind back tool

List FIVE other types of special-purpose tools and equipment:

1 _____

2 _____

3 _____

4 _____

5 _____

INSPECTION AND CARE OF TOOLS AND EQUIPMENT

Before commencing work, the condition of hand tools should be checked and any that are damaged should not be used. Similarly, any special service tool drawn from the store should be inspected to ensure that it is safe to use and capable of doing the work required.

State examples of when hand tools should be discarded or, if possible, repaired:

Spanners: _____

_____

Chisels: _____

Hammers: _____

Screwdrivers: _____

_____

Sockets: _____

Ratchets: _____

Certain test equipment should be checked against known standard readings. Name service items that require testing in such a manner:

_____

_____

_____

_____

_____

_____

_____

⚡ Never use a damaged tool or piece of workshop equipment.

Either safely and correctly repair it, or dispose of it in an environmentally friendly way.

If the damaged tool or equipment cannot be repaired immediately, take it out of service, label it with defect details and ensure it is repaired as soon as possible by a competent person.

How should tools and equipment be kept to ensure they are secured against loss?

_____

_____

_____

What care procedures need to be carried out to ensure tools and equipment are not damaged?

_____

_____

After using tools and equipment what general care procedures need to be carried out?

_____

What checks should be carried out on mains electrical equipment (drills, grinders, etc)?

_____

_____

Insert the missing words in the following paragraphs using the words in the word bank below (note: there are two extra distracter words):

power	over	fuse	overheating
wires	unwound	floor	cooling
damage	secure	roof	

Extension leads must always be completely _____ before connecting to a _____ supply

and using. This prevents _____ of the cable. Check the cable for _____ and exposed

_____. The plug must be _____, undamaged and fitted with the correctly rated

_____.

Never run trolley jacks or other equipment _____ cables or airlines lying across the workshop

_____, as this can damage them.

STORAGE

After purchasing and acquiring the correct tools it is important to store them in a suitable cabinet. The cabinet needs to be lockable and a place to keep the tools clean and dry. Tools should be stored logically so as to be easy to find when required. The most popular type of tool box is the roll cabinet (see figure opposite).

A typical roll cabinet

MEASURING TOOLS

Correctly name the tool opposite.

These can be used to measure _____
_____. They can be
purchased with a digital readout which makes
them easier to use. They have an accuracy of
0.02 mm.

Name one use for this piece of equipment:

_____

Micrometer

The micrometer caliper is used to measure the diameter of components to an accuracy of
0.01 mm (1/100 mm). Internal measuring micrometers are also available.

Micrometers are made in size ranges of 0–25 mm, 25–50 mm, 50–75 mm, etc.

State ONE component that an
external micrometer could be
used to measure:

_____

State ONE component that an
internal micrometer could be
used to measure:

_____

An external micrometer

All above images supplied by Draper Tools Ltd

Dial indicators

Dial test indicator

A dial test indicator (DTI) is an instrument which may be used to give comparative measurements
from an item of specific or standard size.

State ONE component that a dial test indicator can be used to measure:

_____

Identify the measuring tool shown above: _____

What is the difference between a rule and a ruler? _____

_____

Both the rule and the ruler are used to mark out straight lines. They may be calibrated in either
metric, imperial or both measurements.

Feeler gauge

These are a selection of thin strips of metal of specific thickness and have an accuracy of 0.025 mm. They are used to measure the size of clearance gaps. They can be obtained in metric or imperial measurements, or a combination of both.

Give TWO examples of where these would be used:

1 _____

2 _____

Tape measure

These are used to measure length and distances between objects. They generally have a tab on the end which can be hooked over the edge of an object, allowing one person to extend the tape and carry out measurements. The increments are normally in 1 mm graduations. The tape measure can be locked in position and has an automatic spring load return for the tape.

Tape measure

Volume measures

These are used to accurately measure fluids, such as engine oil. They can measure liquids in millilitres and litres.

5 L Measuring jug

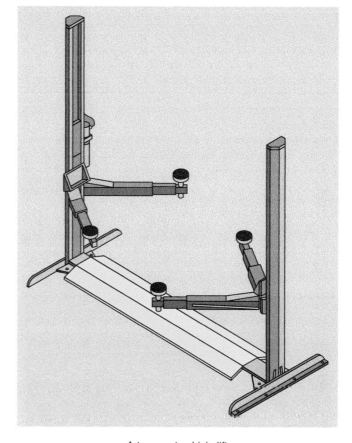

A two-post vehicle lift

Vehicle lifts/hoists generally come in the following types:

- **Four-post**
- **Two-post**
- **Single post**

There are also individual lifting machines which can be linked together, which are used on trucks and buses. These types of lifts are portable and one lift goes under each of the wheels on a four-wheeled vehicle. When they are electronically connected together, the operator can lift all of them at the same time by using a controller. It is usual safe practice to support the vehicle on suitable stands.

What is SWL an abbreviation of and what does it refer to? _____

_____

The hoist should have the SWL displayed on it. If it does not, the hoist needs to be load tested and referred to the manufacturer for the correct SWL.

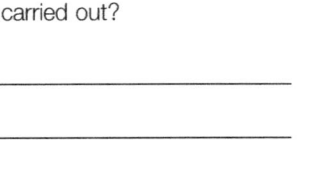

The SWL must NEVER be exceeded on any lifting or weight supporting equipment.

Always check the SWL before using or operating any carrying equipment.

This includes: vehicle hoists, ramp, axle stands, transmission jacks, engine hoists, chains and straps.

Before using a hoist, be certain that you know how to use it safely.

You can find this out by referring to the manufacturer's operating instructions before using a hoist. Ideally you should be suitably trained in how to safely operate it.

What routine checks need to be carried out on vehicle hoists?

- _____
- _____
- _____

When operating a vehicle hoist, what safety precautions need to be carried out?

- _____
- _____
- _____

TROLLEY JACKS AND STANDS

Using a jack to raise a vehicle and supporting it on axle stands

Insert the missing words into the following paragraph using words from the word bank below (note: there are two extra distracter words):

handle	**portable**	**static**	**hydraulic**	**forces**
suitable	**weight**	**aligned**	**axle**	
crushes	**jacking**	**pad**	**pumping**	

Confirm the _____ of the vehicle and the SWL of the trolley jack before commencing the lifting operation. Trolley jacks are _____ and mounted on wheels. The lifting pad is placed under the vehicle _____ to the correct jacking point. A _____ is used to operate the jack by using a _____ action, which _____ hydraulic fluid to a _____ cylinder. This raises the _____ until it comes into contact with the _____ point. Continue operating the handle until the vehicle is at a _____ height. Place the correct _____ stands (check the SWL) under the vehicle.

NEVER attempt to lift a vehicle using a jack which exceeds its SWL. This can damage the seal in the jack causing it to break or collapse, resulting in damage to the vehicle, injury or death.

List FIVE checks that should be carried out on a trolley jack before using it:

1 _____

2 _____

3 _____

4 _____

5 _____

Axle stands are designed to support the weight of a vehicle after it has been suitably raised. They come in various types and are designed for a particular application and should NEVER be used for a job they are not intended for. They should always be used as matched pairs.

Stands will have their SWL displayed on them and should only be used to support loads less than this rating

MARKING OUT

Sometimes simple components need to be made for motor vehicles. This can be done by reproducing lines on the surface of the material to be used. An engineering drawing can be used to show the measurements and component to be manufactured. The process is known as marking out.

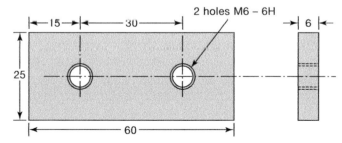

The diagram above shows an engineering drawing for a simple mounting plate. In order to reproduce this on a metal plate, the dimensions need to be transferred to its surface.

What needs to be applied to the surface so marking out can take place? _____

Once this has been applied accurately, dimensioned lines, centre lines, curves and in some cases circles need to be drawn on the surface.

There are a selection of tools and equipment designed to make these tasks easier.

Name the tool/equipment used to carry out the following:

Used to check the flatness of a component: _____

Used to draw lines on the prepared surface: _____

Allows a straight line to be drawn exactly 90° to the edge to be drawn: _____

List EIGHT other tools/equipment which can be used in marking out:

1 _____

2 _____

3 _____

4 _____

5 _____

6 _____

7 _____

8 _____

BENCH VICE

These are used for securely holding pieces of work that are being worked on.

Detachable vice clamps are used to protect the work as the vice jaws are generally rough and made of steel. These are made of some soft metal such as

_____.

Picture supplied by Draper Tools Ltd

Bench vice

THREAD FORMING

During routine maintenance internal and external threads may become damaged, requiring them to be effectively repaired. Sometimes brackets, tools and components need to be manufactured locally in the workshop and these may also need to have threads cut for bolts or studs.

The tools below are used when forming threads are made from high carbon steel. Correctly name these tools:

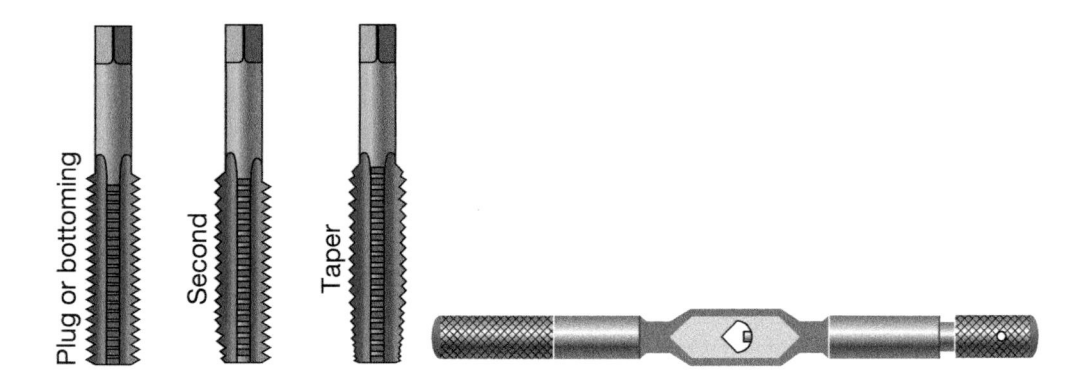

_____   _____

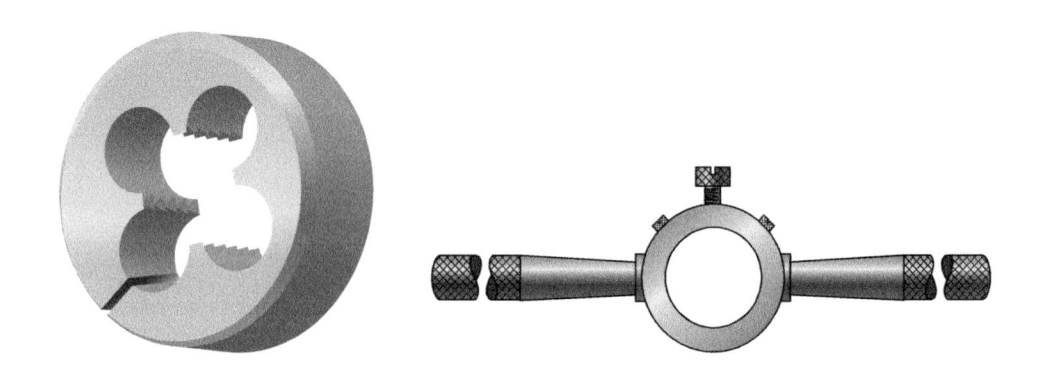

_____   _____

Cutting an internal thread

Insert the missing words into the following paragraphs using words from the word bank below (note: there are two extra distracter words):

bar	aluminium	reversed	brass	second
friction	plug	cut	lubricant	removed
alignment	chips	taper	wrench	

The hole is first tapped with the _____ tap which is held in a _____ or chuck type tap _____. The tap must be kept in _____ with the hole. The tap will start to _____ and as a rule after one complete turn, it will need to be _____ slightly, as this will break away the _____ of metal. This tap is then followed by the _____ tap and then the _____, using the same procedure as with the taper tap.

When tapping a hole it is advisable to use a _____ on all metals, except cast iron and _____, as this reduces _____.

When tapping a blind hole a plug tap must be used last.

Cutting an external thread

When cutting external threads a die and die holder are used.

What is the most popular die that is used? _____

When making the first cut with this type of die, the centre screw has to be screwed all of the way in to open the split.

How should the end of the rod be prepared before using the die to cut the thread? _____

_____

Briefly describe how to cut an external thread with a die:

_____

_____

When the first cut is completed try the threaded rod in the tapper or use a nut. If it is too tight make another cut with the die, first slacken the inner screw and slightly tighten the outer screws.

Lubricants for threading

A suitable lubricant will need to be used when cutting external and internal threads.

Complete the following table:

Material to be threaded	Lubricant
Aluminium	_____
Brass	_____
Bronze and copper	_____
Cast iron	_____
Mild steel	_____

MATERIALS USED IN VEHICLE CONSTRUCTION

There is a vast array of materials used in the construction of the modern motor vehicle. Most of the main mechanical and structural components of a car are made from metal in one form or another.

Complete the figure below using the following list of typical materials, used in the construction of a car:

cast iron   tempered glass   synthetic rubber   pressed steel
aluminium   laminated glass   mild steel

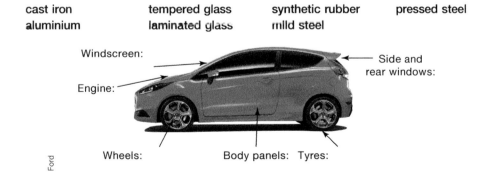

Windscreen:
Engine:
Wheels:
Body panels: Tyres:
Side and rear windows:

Ford

Ferrous and non-ferrous metals

Insert the missing words into the following paragraph using words from the word bank below (note: there are three extra distracter words):

alloyed	components	ferrous	hard
magnetic	soft	not	
less	iron	more	

Metals may be split into two main groups: Ferrous and non-ferrous metals.

A metal that contains iron is _____ and it is also _____. Non-ferrous metals do not contain _____, they are _____ magnetic and tend to be _____ weak materials.

Most non-ferrous metals are _____ (mixed) with other metals, making them _____ useful in the production of motor vehicle _____.

Give TWO examples of ferrous metals:

_____

Give TWO examples of non-ferrous metals:

_____

Plastics

Plastics can be formed into any shape when heat and pressure is applied and are from a large group of man-made materials.

The two types of plastics which are used are:

1 _____

2 _____

Glass Reinforced Plastic (GRP) and carbon fibre are sometimes used in the construction of motor vehicles. Give one likely part of the vehicle where they could be used: _____.

JOINING OF MATERIALS AND COMPONENTS

Various methods are used to join parts and materials which are used in the construction of vehicles. These can include the joining of body, engine, gearbox and other assembled components.

Mechanical joining devices

These provide the means of joining one component to another and are traditionally nuts, bolts, screws, keys and pins and on modern vehicles the use of adhesives is becoming ever increasingly common.

Complete the following table which shows a number of methods of joining vehicle components:

Method of joining	Components
Riveting	
Adhesive bonding	
Shrinking	
Dowel	
Key (e.g. woodruff)	

Locking devices

Mechanical locking devices are commonly used in the construction, repair and routine maintenance of vehicles.

Name the types of locking devices shown below:

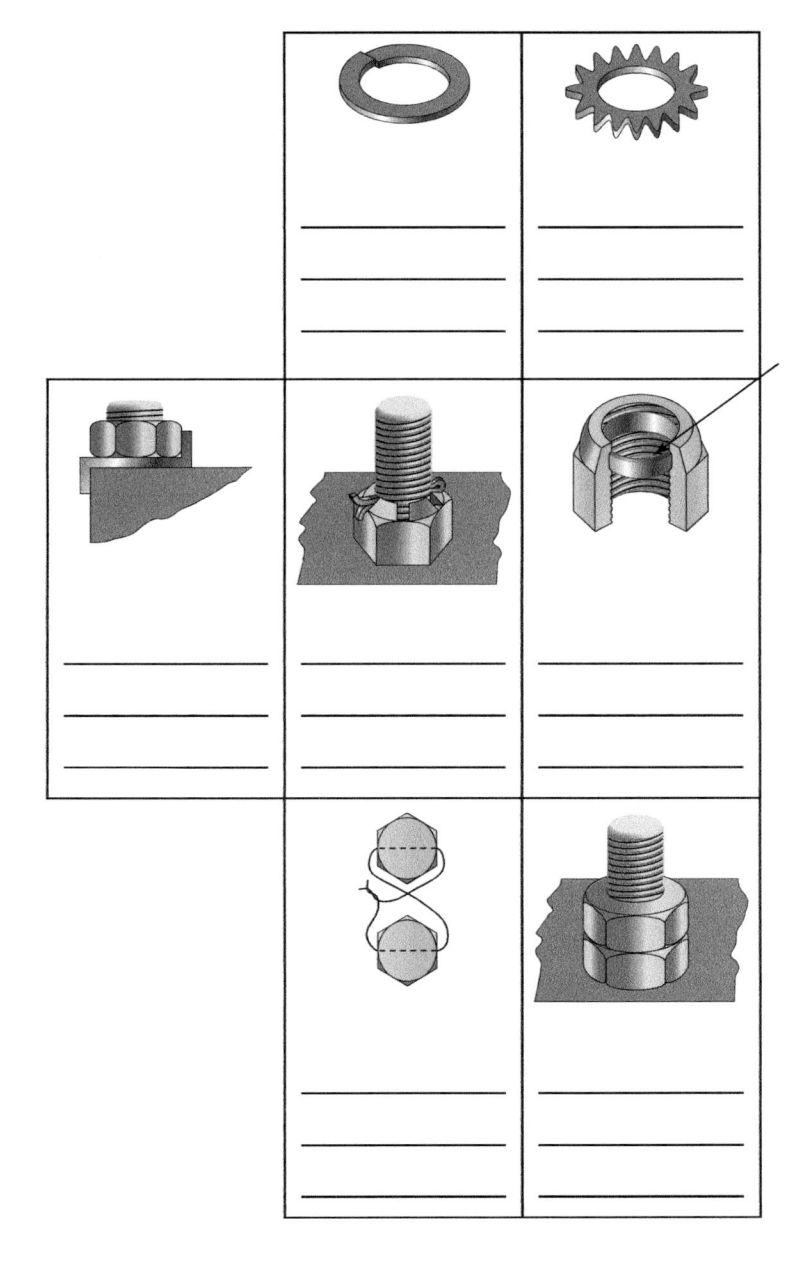

MATERIALS USED DURING MAINTENANCE

During vehicle maintenance a selection of common materials will be used. Complete the following table giving some examples of the uses for each material:

Material	Examples of uses
Lubricants	
Coolants	
Fluids	
Filters	
Gaskets	
Aerosol sprays	
Adhesives and cements	
Sealers	
Cleaners	

Always read the safety instructions on containers and packaging before use.

Explain what components the product shown below is used on?

_____

Comma Oil

ACROSS

1 This is straight and calibrated from the very end.

2 A spanner which has an open jaw and ring at each end is called a _____ spanner.

3 Before using a hammer the head and shaft must be checked for _____.

6 A type of coarse file.

7 A hacksaw blade should be fitted with the teeth facing _____.

8 It is hit with a hammer and used to cut through metal.

9 PPE which must be worn when drilling.

DOWN

1 This is a material used in the construction of a tyre.

2 Tyre pressure gauges and torque wrenches need to be regularly _____.

3 What will be weakened if a torque wrench is not wound off?

4 Ferrous metals contain this.

5 This type of caliper can be used for measuring.

Multiple choice questions

Choose the correct answer from a), b) or c) and place a tick [✓] after your answer.

1 **SWL is an abbreviation of:**

  a) Safe Working Limit [ ]

  b) Safe Working Load [ ]

  c) Special Working Limit [ ]

2 **What lubricant needs to be used when cutting a thread in brass?**

  a) Paraffin [ ]

  b) None [ ]

  c) Gear oil [ ]

3 **Which one of the following is a locking device?**

  a) Nylock [ ]

  b) Tap [ ]

  c) Micrometer [ ]

SECTION 5

Introduction to the retail automotive maintenance and repair industry

USE THIS SPACE FOR LEARNER NOTES

_____

_____

_____

_____

_____

Learning objectives

After studying this section you should be able to:

- Identify different types of vehicle in the automotive sector.
- Describe the features of different car types.
- Describe types of organizations found in the automotive sector.
- Recognize types of job roles in the automotive sector.
- Describe the type of work carried out in a range of job roles.
- Recognize an automotive organizational structure.

Key terms

Franchised dealer This is a garage dealership linked to a vehicle manufacturer and can be known as a main dealer.

After sales (service) department Section of a garage dealing with routine servicing, diagnosis and repair of faults and often MOT tests.

Parts department Section of a dealership selling spare parts to trade and retail customers.

Organizational structure Plan showing the relationship between job roles and lines of authority (who is responsible to who).

Job description Document that gives specific details of a job role.

VEHICLE TYPES

There are numerous types of vehicle found in the automotive sector, which are used for different purposes.

Label the vehicle types in the following pictures using the terms listed below:

Saloon car	Motorcycle
Estate car	Coach
Convertible	Car derived van
3 door hatchback	Rigid truck
5 door hatchback	Tractor unit
Coupe	Minibus
Moped	Tractor and trailer unit
Multipurpose vehicle (MPV)	Panel van
4 × 4 Vehicle	Articulated tractor and trailer unit

Mercedes

Mercedes

_____

Vauxhall

_____

Vauxhall

_____

Ford

_____

BMW

_____

Ford

_____

Mercedes

Ford

Vauxhall

Peugeot

Mercedes

BMW

Suzuki

Discuss in groups what each of the vehicles shown in the pictures will be used for.

What are the main features and positive aspects of each vehicle?

Use the internet to help you.

Draw lines to match the distinguishing features of the following car types:

Convertible	Car with extended boot space	Saloon car
	Car with lockable boot and fixed parcel shelf	
Estate car	Car where the roof can be folded back	Hatchback
	Sporty car usually with only two seats	
	Car without rear side windows, two front seats and doors at the rear	
MPV	Car with a tailgate and folding rear seats	4 x 4
	Car designed for off-road use	
Coupe	Car that can carry up to seven people, with the option to remove all rear seats	Car derived van

AUTOMOTIVE ORGANIZATION TYPES

The retail motor industry is made up of many different types of organizations, covering a wide range of services to road using customers.

WWW   **www.autocity.org.uk** – select world of work

Using the 'Autocity' website identify nine types of organizations.

No.	Type of organization
1	
2	
3	
4	
5	
6	
7	
8	
9	

Select THREE of the organizations on page 54 and state the type of work that would be carried out.

1 Type of organization: _____
   Type of work carried out:
   - _____
   - _____
   - _____
   - _____

2 Type of organization: _____
   Type of work carried out:
   - _____
   - _____
   - _____
   - _____

3 Type of organization: _____
   Type of work carried out:
   - _____
   - _____
   - _____

DEALER TYPES

The terms used to describe the activities of different dealership organizations are:

- **Franchised dealer**
- **Multi-franchised dealer**
- **Non-franchised dealer**

Match the following statements to the terms opposite:

A _____ dealer sells new and used vehicles, services, repairs and supplies spare parts for vehicles of a specific manufacturer. The dealer is linked to the manufacturer using common branding, e.g. Honda.

A _____ dealer may sell some used cars and services and repairs a wide range of vehicles. This dealer is known as an independent and is not linked to a vehicle manufacturer.

A _____ dealer has more than one franchise on their site. The dealerships usually are stand alone and the showrooms, workshops and parts departments are not shared with other franchises.

Use local newspapers or the Internet to find examples of each of the above types of dealerships in your area, and record the details below.

Some may have websites to find information about the services offered.

Dealership type:_____

Name_____

Details_____
_____

Dealership type:_____

Name_____

Details_____
_____

Dealership type:_____

Name_____

Details_____
_____

TIP Yellow pages or **www.yell.com** would be a good starting point.

JOB ROLES WITHIN THE RETAIL MOTOR INDUSTRY

Using the autocity website on page 54, research the following job roles and list THREE duties involved within that job:

Car dealership, independent garage and parts factor

Light vehicle technician

Type of work:

- _____
- _____
- _____

Customer service advisor (service receptionist)

Type of work:

- _____
  _____
- _____
- _____

Parts advisor

Type of work:

- _____
- _____
- _____

Vehicle sales advisor (salesperson)

Type of work:

- _____
  _____
- _____
- _____

Fast fit centre

Fast fit technician

Type of work:

- _____
- _____
- _____

Tyre technician

Type of work:

- _____
- _____
- _____

Body repair centre

Paint technician

Type of work:

- _____
- _____
- _____

Panel technician

Type of work:

- _____

  _____

- _____

- _____

AUTOMOTIVE COMPANY ORGANIZATIONAL STRUCTURE

The organizational structure indicates who does what in the company. It explains each person's duties and level of authority. It also shows everyone's working relationships, from the most senior manager to the most junior employee.

There are many automotive companies in the United Kingdom, ranging from sole owner garages to large corporate organizations. The organizational structure activity opposite relates to a large franchised dealership that sells and repairs cars.

Use the job roles below to complete the organizational chart showing the lines of authority.

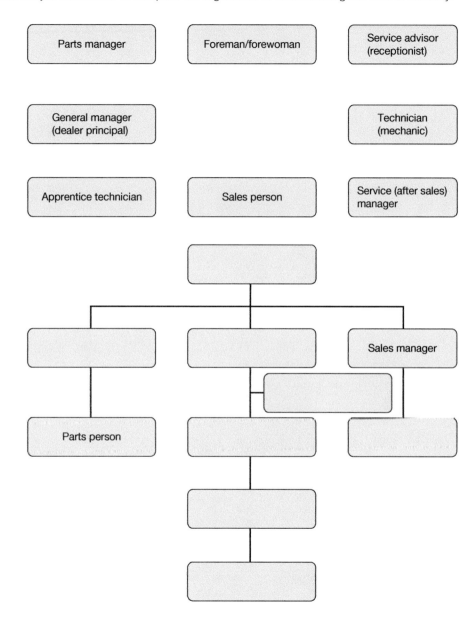

Multiple choice questions

Choose the correct answer from a), b) or c) and place a tick [✓] after your answer.

1 Which type of vehicle is likely to carry up to seven people?

a) 3 door hatchback [ ]

b) Coupe [ ]

c) Multipurpose vehicle (MPV) [ ]

2 What type of vehicle is shown?

a) 5 door hatchback [ ]

b) 4 door saloon [ ]

c) Estate car [ ]

Ford

3 Which type of vehicle has a removable roof?

a) 4 × 4 [ ]

b) Convertible [ ]

c) Hatchback [ ]

4 Fast fit technicians are likely to carry out which of the following tasks?

a) MOT test [ ]

b) Rebuilding automatic gearboxes [ ]

c) Replacing tyres [ ]

5 Which type of organization is likely to service and repair a wide range of vehicle makes?

a) Franchised dealer [ ]

b) Independent dealer [ ]

c) Body repair centre [ ]

PART 2
ENGINE CONSTRUCTION AND OPERATING PRINCIPLES

USE THIS SPACE FOR LEARNER NOTES

SECTION 1
Engine construction 61

SECTION 2
Engine liquid cooling and lubrication systems components and operation 76

SECTION 1

Engine construction

USE THIS SPACE FOR LEARNER NOTES

_____

_____

_____

_____

_____

_____

Learning objectives

After studying this section you should be able to:

- Work safely on SI and CI engines.
- Identify SI and CI engine system components.
- State how SI and CI engines operate.
- State the purpose of the main systems associated with engines.

Key terms

Crankshaft Bolted into the lower part of the engine block. It converts linear motion into rotational motion.

Firing order This is the order that combustion takes place in multi-cylinder engines.

Camshaft Rotates in the engine and opens the valves at the correct time.

Valve timing The correct time when the valves open and close.

Four stroke The operating cycles for complete and full combustion in the Otto cycle.

www.animatedengines.com

www.familycar.com/Engine.htm

When working on vehicles always wear the correct PPE: safety footwear, overalls, goggles, safety gloves and ear defenders.

```
L R I N D U C T I O N
O E N D E K O R T S C
N D P R V I M G P B A
A N P C L M P C O D P
G I O D A O R R M H A
N L W B V M E L C W C
F Y E T E K S A G D I
E C R A N K S H A F T
V N N O T S I P A R Y
V M A N I F O L D F D
R N S O P C N R M I T
```

GASKET	MANIFOLD	CYLINDER
PISTON	VALVE	INDUCTION
CRANKSHAFT	COMPRESSION	CAPACITY
POWER	CAMSHAFT	BORE
BDC	TDC	STROKE

 **RMP** When working on engines (as with all vehicle systems) refer to the vehicle manufacturers' repair instructions for the torque settings, fluid types and quantities.

When working on a vehicle's engine, as with working on any other vehicle system, it is important to protect the interior and exterior of the vehicle. This is achieved by using the correct PPE (Personal Protective Equipment) and VPE (which is an abbreviation for _____ _____).

When working on customers' vehicles it is important to protect the interior. Name two pieces of VPE which should be used inside a vehicle:

1 _____

2 _____

It is also important to protect the exterior of the car, especially when working in the engine compartment. Name the VPE which should be used in this area: _____ .

What are these pieces of VPE protecting the vehicle from?

_____

_____

ENGINES

The engine of a motor vehicle provides the _____ to propel the vehicle and to operate the various ancillaries, such as the alternator, water pump, vacuum-assisted brakes and air conditioning. However, when the accelerator is released and the road wheels are made to turn the engine (when fitted with a manual gearbox), it provides a useful amount of vehicle _____ without the use of the brakes.

When talking about engines we normally refer to them as SI and CI engines.

What does SI stand for? _____

What does CI stand for? _____

ENGINE TYPES

Engines used in motor vehicles may be referred to as four-stroke or two-stroke engines.

The term 'stroke' refers to: _____.

A four-stroke engine is: _____
_____

A two-stroke engine is: _____
_____

The number of revolutions completed during a working cycle on a four-stroke engine is: _____

The number of revolutions completed during a working cycle on a two-stroke engine is: _____

Complete the table to identify different types of engines used in motor vehicles.

This can be carried out individually or in pairs. Use a combination of technical data books, electronic technical data and real vehicles in your workshop.

Vehicle	No. of cylinders	Type of fuel	Engine capacity

ENGINE CONSTRUCTION (SI AND CI)

The engine consists of a number of components. When all of these components are correctly fitted together they allow the fuel and air to combust (explode), which is turned into rotational power that goes out to the wheels.

We will now look at the main components.

Always dispose of waste oil, filters, rags and old engine components correctly.

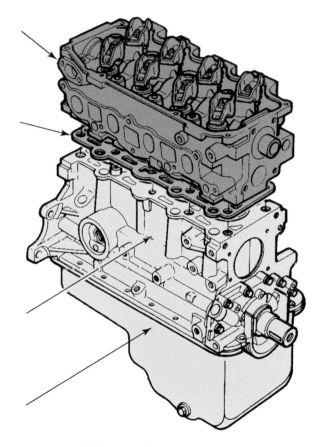

Major units of an engine

Label the components on the diagram on page 63 using the terms listed in the table below. State their basic function or purpose in the table.

Component	Function/purpose
Cylinder block	
Engine sump	
Cylinder head	
Head gasket	

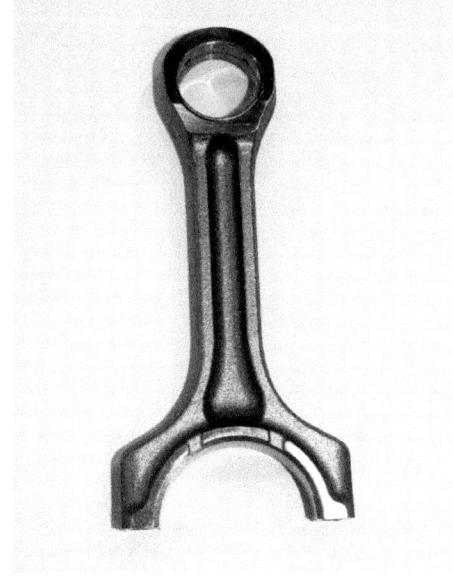

We have had a basic look at the major components. Now identify and correctly label the following component images:

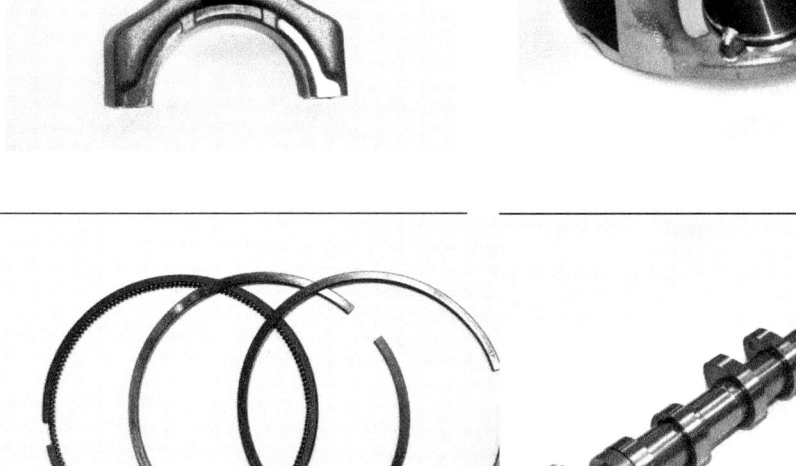

Correctly label the above image with the following:

web	**counterweight**	**mounting**
main journal	**flywheel**	**flange**

Piston

The piston fits securely inside the cylinder in the cylinder block. It is made of a lightweight metal generally an aluminium alloy.

Briefly explain what happens to the piston as combustion takes place above it:

_____

_____

Piston rings

These seal around the piston, preventing gases from passing by and ensuring a gas tight fit.

There are normally three rings fitted to a piston. The top two rings help to seal the piston and prevent the pressure above from escaping. What are these rings called?

_____

What is the purpose of the bottom ring?

_____

Sketch a piston. Correctly label it with the following terms:

Piston crown   **Compression rings**   **Gudgeon pin hole**
Skirt   **Oil control ring**

Insert the missing words into the following paragraphs using the word list below (note: there are two extra distracter words):

small	open	con	alloy	upper
crankshaft	casting	half	plastic	rotational
linear	profile	lobes	big	valves
crankpin	strength	gudgeon	lower	

Connecting rod

This is also known as the '_____ rod'. The top end is known as the _____ end and is

connected to the piston by the _____ pin. The bottom end is called the _____ end and is

connected to the _____ by a bearing surface. The **crankshaft** is made from a one piece

_____, or forging, of heat-treated _____ steel of great mechanical _____.

Crankshaft

The connecting rod bottom end, known as the big end, is attached to the _____ crankpin. It is mounted in the _____ end of the cylinder block where it rotates, providing drive from the engine. The crankshaft turns the _____ (up and down) motion of the piston into _____ (round and round) motion.

Camshaft

This rotates and is designed to _____ the inlet and exhaust _____ at the correct time. The **camshaft** rotates at _____ the speed of the crankshaft.

The camshaft is a metal shaft with machined '____'. These lobes make contact with a mechanism which opens the valve. The _____ (shape) of the lobe determines how much and for how long the valve opens. There could be one or two camshafts fitted to an engine.

Valves

These are mounted in the cylinder head. There is normally one inlet and one exhaust valve per cylinder. A lot of modern cars can have more valves fitted than this.

What is the purpose of valves?

_____

_____

Valve springs

A valve spring being measured

Valves need to be closed quickly and have a gas tight seal. This is done by the valve spring.

Inlet and exhaust manifolds

These direct the air (and fuel for petrol engines) into the cylinders and allow the exhaust gases to safely leave the cylinders to the exhaust system. The internal surfaces of the manifolds are clear and smooth to allow the air and exhaust gases to take an unrestricted route.

What does this design feature help to achieve? _____

Flywheel

Insert the missing words into the following paragraph (note: there are two extra distracter words):

crankshaft	rotating	engages	inner
mass	idle	ring	ignition
clutch	rotate	outer	disengages

The flywheel is a large heavy metal _____ that is bolted to the end of the _____. It helps

to keep the engine _____ during the _____ strokes of the engines cycles. It also serves

as mounting face for the _____ assembly and has the starter _____ gear (a large toothed

gear) securely fitted to its _____ edge. When the starter motor is operated by the _____

key it _____ with this ring to _____ the crankshaft and start the engine.

Front drive pulley

This is mounted on the opposite end of the crankshaft to the flywheel. It allows a drive belt to be fitted and transfer power to the ancillaries.

List FOUR ancillaries driven by the front drive pulley:

1 _____

2 _____

3 _____

4 _____

Gaskets and seals

The engine is made up of a number of components that are joined together. To prevent loss of fluids (oil, petrol, diesel, coolant) and gases (exhaust, air), due to leakage, gaskets are fitted between the contact surfaces. These gaskets are made from thin materials.

List FOUR gaskets used in the construction of an engine:

1 _____

2 _____

3 _____

4 _____

 TIP After any mechanical work, especially engine rebuilds, the engine crankshaft must be rotated by hand. This is to make sure it turns over freely.

On shafts which rotate, a lip type seal is used to prevent the loss of lubricating oil.

Name TWO shafts which would have lip type seals fitted to them:

1 _____

2 _____

Lip type seal

 RMP Check for any fluid leak from seals, gasket, pipes hoses and gasket joints. Report any leaks to your supervisor and customer. Check, clean and rectify any leaks as appropriate. If left unattended leaks can cause serious damage due to loss of fluids or contamination of components.

Turbocharger

A turbocharger

Insert the missing words into the following paragraph (note: there are two extra distracter words):

aspirated	down	exhaust	performance
pressurized	up	impeller	turbine
forced	air	combustion	vacuumed

An engine which does not have a turbocharger fitted to it is known as a normally _____

engine. This means that _____ (at atmospheric pressure) normally enters the engine as the

piston moves _____ the cylinder during the four-stroke cycle. To help improve the

_____ (torque, power) and fuel consumption of engines, air is sometimes forced into the

_____ chamber by a turbocharger. This is known as _____ induction. The turbo-

charger is driven by the _____ gases that leave the engine. These gases turn a

_____ that is connected by a shaft to an impeller at the other end. The _____ forces

air into the engine. In the figure above the red represents the flow of exhaust and blue shows the

intake air that is _____.

A turbocharger is fitted to some petrol engines and is a common feature on diesel engines.

DIFFERENCES BETWEEN DIESEL AND PETROL ENGINE CONSTRUCTION

Complete the table below to show the difference between the engine components in a petrol or diesel engine.

Component	Diesel	Petrol
Cylinder head		
Cylinder block		
Crankshaft		
Connecting rod		
Piston rings		

STARTING SYSTEM

The engine needs to rotate at a high speed to start it. This is done by means of a starter motor. The driver turns an ignition key which operates the starter motor. A gear rotates on the end of the starter motor, which engages with the starter ring gear on the flywheel. This rotates the crankshaft until the engine starts, then the starter motor will disengage. The starter motor needs to be powerful, especially with diesel engines, as they have what is called high compression ratios.

CHARGING SYSTEMS

Vehicles require electrical energy to start them and to operate the various electrical systems they have, such as the lighting system. A battery is fitted to store electrical energy and if it is not continually charged up it will lose its charge and eventually go flat.

To keep the battery charged up, a generator is fitted. This is called an alternator. The alternator is normally driven by a belt from the crank pulley.

When handling old engine oil always wear protective gloves. This is because it is carcinogenic (cancer forming).

LUBRICATION SYSTEM

The engine has a lot of moving parts which need to be kept lubricated so that they move freely without rubbing together. This is achieved by the lubrication system. Engine oil which is stored in the sump is pumped and filtered around to the various parts of the engine.

As well as lubricating the engine, what other function does the oil have?

_____

Whenever you finish working on an engine start it and check for leaks.

State the main purpose of the following engine systems:

Induction: _____

Fuel: _____

Exhaust: _____

Cooling: _____

_____

Ignition: _____

Always check the engine oil when the vehicle is on a flat surface and the engine is cold. The oil level mark on the dipstick should be between the minimum and maximum marks.

ENGINE TERMINOLOGY

When talking about engines there are certain terminologies (words and abbreviations to do with engines) which are used.

Draw lines from the terminology to their correct meaning:

Terminology
Bore
BDC
Stroke
TDC

Meaning
Top Dead Centre. This is when the piston is at the top of its stroke.
The distance the piston moves from the top to the bottom of the cylinder.
Bottom Dead Centre. This is when the piston is at the bottom of its stroke.
The diameter of the cylinder.

On the diagram correctly label the following:

Bore   **Stroke**   **TDC**   **BDC**

Swept volume

This is the volume above the piston crown as it travels from BDC to TDC, as shown by the shaded area in the diagram.

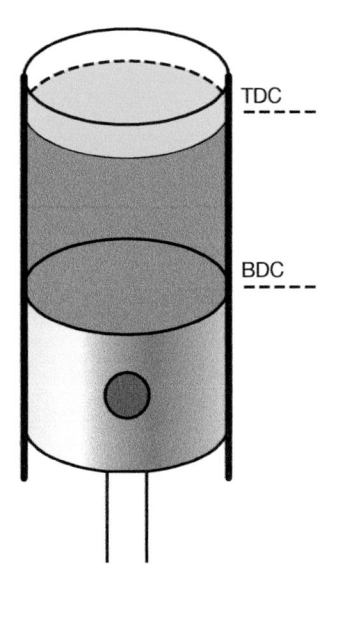

TDC

BDC

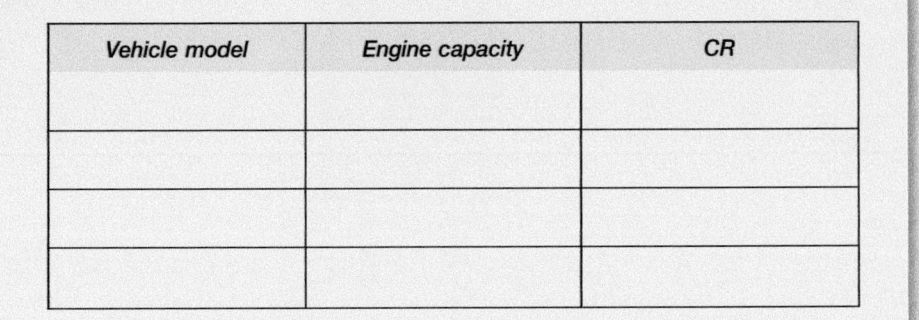

Examine data books to find the compression ratio (CR) of FOUR types of different engines:

Vehicle model	Engine capacity	CR

Capacity

When we talk about engines it is common to mention engine 'capacity'. This tells us the engine size.

How is the total capacity of a 4-cylinder engine calculated?

_____

_____

Compression ratio

This is the ratio of the complete volume of air and fuel in the cylinder, when the piston is at BDC, compared to the volume that it is compressed into when the piston is at TDC (clearance volume).

FIRING ORDERS FOR 4-CYLINDER ENGINES

The most common way of arranging cylinders in an engine is in-line.

Single cylinder

In a single cylinder engine all four strokes take place one after the other.

Twin cylinder

This is very much like having two single cylinder engines joined side by side. Each piston will be on a different stroke in the cycle as this helps to balance the engine and make it run smoothly. See the image at the top of page 71. The cranks may be arranged in two ways, either as parallel cranks (Figure a) or cranks 180° apart (Figure b).

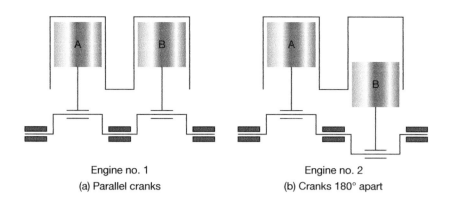

Engine no. 1
(a) Parallel cranks

Engine no. 2
(b) Cranks 180° apart

Referring to the above engines	Engine 1.	Engine 2.
What stroke will piston B be on when A is on the power stroke?		
How many crankshaft degrees will the intervals be between the power impulses?		

4 Cylinder

In multi-cylinder engines, such as a 4-cylinder engine, each cylinder will be in a different part of the cycle.

Complete the line diagram to show a 4-cylinder engine:

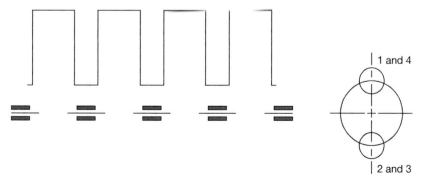

| 1 and 4

| 2 and 3

When the figure below left is correctly completed it will show the positions of the pistons in a typical 4-cylinder engine. Piston number '1' is at TDC on the compression (ignition) stroke. As the crankshaft continues to rotate, the next piston to reach TDC on the compression stroke is '3', then '4' and finally '2'. This gives a firing order of 1342.

Give ONE other common firing order for a four-stroke in-line 4-cylinder engine?

_____

For the fuel and air to ignite, then combust, a spark needs to occur at the correct time. This spark will occur when the piston is on which stroke? _____

As the engine speed increases the spark will need to occur earlier for the fuel/air to combust in a suitable timescale. This is known as varying the _____

FOUR-STROKE CYCLE – SPARK IGNITION (SI)

The four-stroke cycle is also known as the Otto cycle, named after the first person to build it in 1876, a German named Nicolaus Otto.

The spark ignition engine normally uses petrol as a fuel, which when mixed with air is ignited by a spark at the correct time in the engine operating cycle.

The four-stroke cycle is completed in four movements of the piston during which the crankshaft rotates twice.

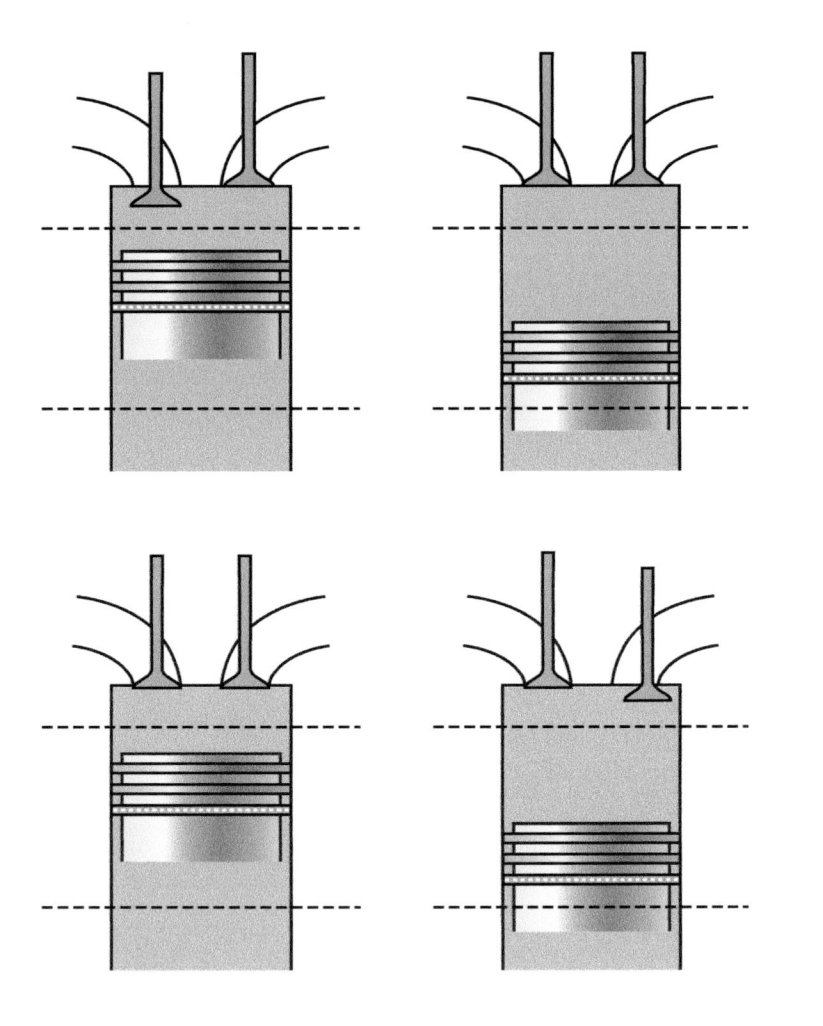

Complete the line diagrams above to show the positions of the valves at the start of each stroke. Name the other three strokes. Indicate the direction in which the piston is moving in each case.

Insert the missing words into the following paragraphs using the word list below (note: there are two extra distracter words):

open	upwards	closed	lowering
compressed	air	top	spark
inlet	up	gases	closed
ignites	down	crankshaft	open
downwards	fuel	rising	

Induction

The _____ valve is _____ and the piston is moving _____ the cylinder; a mixture of _____ and _____ is drawn into the cylinder.

Compression

Both valves are _____ and the piston moves _____ the cylinder. The mixture is then _____ until the piston reaches the _____ of the stroke, just before this point a _____ _____ the mixture.

Power

Both valves are _____ as the piston is forced _____ by the rapidly burning and strongly expanding _____. This produces the force to rotate the _____ and produce power.

Exhaust

The exhaust valve is _____ allowing the still pressurized burning gases to rush out. The _____ piston clears out the remaining spent gases.

COMPRESSION-IGNITION (CI) DIESEL ENGINE – FOUR-STROKE CYCLE

The actual strokes, induction, compression, power and exhaust, are exactly the same as in the SI engine and it normally uses diesel as fuel. The operating principle is slightly different to that of the SI engine, as described below.

Insert the missing words into the following text using the word bank below (note: there are two extra distracter words):

closes	gases	compressed	pressure
opens	spray	injected	rising
air	inlet	temperature	liquid

Induction: The piston is descending and with the _____ valve open _____ only enters the cylinder.

Compression: When the inlet valve _____ the air is _____ to between 16:1 and 24:1. This causes the air _____ to rise substantially. Approximately 15–18° BTDC fuel is _____ into the cylinder in the form of a fine _____; it mixes rapidly with the air and the high temperature causes combustion to take place.

Power: Combustion causes a rapid _____ rise and the piston is _____ downwards.

Exhaust: The _____ piston forces the burnt _____ out of the cylinder.

List the main differences of the compression-ignition engine when compared with the spark-ignition engine.

1 _____
2 _____
3 _____
4 _____
5 _____

TWO-STROKE PETROL ENGINE CYCLE (CRANKCASE COMPRESSION TYPE)

By making use of both sides of the piston, the four phases, induction, compression, power, exhaust, are completed in two strokes of the piston or one crankshaft revolution.

The mixture entering and leaving the combustion chamber is controlled by the piston. It acts as a valve covering and uncovering ports in the cylinder wall. Induction into the crankcase can be controlled by the piston (via the piston port) by reed valves or by a rotary valve.

As the piston rises it creates a partial vacuum in the sealed crankcase. This causes the mixture to be drawn into the crankcase, either through the reed or rotary valve.

The mixture transfers from the crankcase to the combustion chamber. As the piston moves down the cylinder, the inlet port or valve closes. Pressure builds up in the sealed crankcase as space is reduced. When the piston uncovers transfer ports, the mixture is forced into the combustion chamber.

Explain how a two-stroke engine is lubricated: _____

_____

Name the main parts and show the direction of crankshaft rotation on the diagram below.

(a) **Show fuel entering the crankcase.**
(b) **Show fuel transferring to the cylinder and exhausting.**
(c) **Indicate the reed valve on the alternative intake system.**

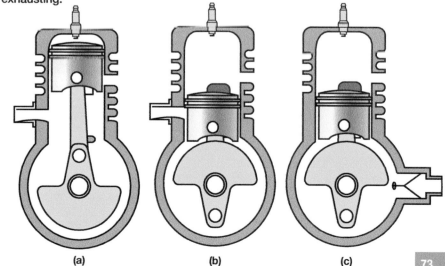

(a)   (b)   (c)

SUMMARY OF TWO-STROKE CYCLE

Complete the table to indicate what is happening above and below the piston on each of the strokes:

Events	Stroke upward	Stroke downward
Above the piston	Closing of transfer port. Completion of exhaust. Compression.	_____ _____ _____ _____
Below the piston	_____ _____	Partial compression of new mixture in crankcase.

VALVE TIMING

Fuel and air must get into the engine very quickly, completely burn and the exhaust gases leave the engine fast. Because this happens in such a short space of time in the four-stroke cycle, the valves, under certain operations, will need to open early and close late. This applies to both four-stroke SI and CI engines.

The timing of the valve movements to the position of the piston is known as valve timing.

When a valve opens early it is known as _____

Where a valve closes late it known as _____

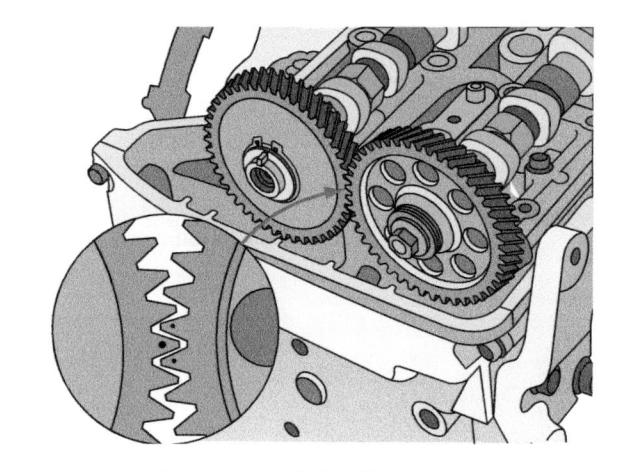

A typical valve timing diagram

Engine designers make use of the valve timing to allow the incoming air and fuel (CI air only) to help force the exhaust gases out. The exhaust gases which are exiting also help to draw the new mixture into the cylinder. At this point both valves will be open.

This is known as _____

IGNITION TIMING

The ignition system provides the spark at the correct time to start combustion

In an SI engine the spark needs to happen at the correct time.

Why is this? _____

_____

As the engine runs faster the spark will need to happen earlier to allow the fuel/air mixture time to burn.

This is called _____

In a CI engine there is no spark. The fuel that is injected into the cylinder needs to be injected earlier as the engine speed increases, just the same as the SI engine.

This is called _____

Where the diesel fuel is injected into the combustion chamber above the piston it is known as _____. When the diesel fuel is injected into a small area of the cylinder head called the swirl chamber, it is known as _____

EXHAUST EMISSION

Dirty exhaust emissions are bad for the environment and health

During the operation of a four-stroke engine, combustion (burning) of the fuel and air takes place. A number of gases will pass out of the exhaust system and into the atmosphere.

In a CI engine the following gases are emitted. Circle the ones which are a concern for the environment and health:

H_2O	N	CO	NOx
O	CO_2	HC	particulates

The SI engine emits the following gases. Circle the gases which are a concern to the environment and health:

H_2O	N	CO	NOx
O	CO_2	HC	

Multiple choice questions

Choose the correct answer from a), b) or c) and place a tick [✓] after your answer.

1 **Which one of the following closes the valves?**

a) Camshaft [ ]

b) Piston crown [ ]

c) Spring [ ]

2 **When a valve opens early it is known as:**

a) Valve lag [ ]

b) Valve lead [ ]

c) Valve lag [ ]

3 **A typical firing order for a 4-cylinder engine is:**

a) 1342 [ ]

b) 1234 [ ]

c) 1432 [ ]

4 **The flywheel is bolted to the:**

a) Camshaft [ ]

b) Engine block [ ]

c) Crankshaft [ ]

5 **Compared to the crankshaft the camshaft rotates at:**

a) Half the speed [ ]

b) Double the speed [ ]

c) The same speed [ ]

Engine liquid cooling and lubrication systems and components and operation

_____

_____

_____

Learning objectives

After studying this section you should be able to:

- Work safely on engine liquid cooling and lubrication systems.
- Understand engine liquid cooling and lubrication systems components.
- Understand how engine liquid cooling and lubrication systems operate.
- Carry out routine maintenance on engine liquid cooling systems.
- Carry out routine maintenance on engine lubrication systems.

Key terms

Conduction When heat passes through solid materials, mainly metals.

Convection When heat is carried by moving liquid or gas in an upwards direction.

Radiation When heat is given off into the air, from the surface of the object.

Radiator Coolant flows through it and transfers heat from the engine to the surrounding air.

Thermostat Temperature sensitive valve that controls coolant flow to the radiator.

Pressure cap Maintains the correct operating pressure of the cooling system.

Water pump Circulates coolant around the engine.

Antifreeze Added to water to lower freezing point and protect the engine from corrosion.

Hydrometer Equipment used to test the specific gravity of a liquid.

Drive belt/serpentine belt Commonly referred to as the fan belt. This can be used to drive the water pump, alternator, power steering pump and some auxiliary units.

Cooling fan Helps to cool the radiator and maintain the coolant temperature under extreme conditions.

Expansion tank Fitted to some cooling systems as a reservoir.

www.gm-radiator.com

www.commaoil.com

When working with engine cooling systems you may be dealing with:

- Very hot surfaces
- Very hot oil or coolant under pressure, even when draining the sump
- Chemicals which are carcinogenic or toxic
- Rotating components
- Waste disposal

Hazardous substances

Many liquids and substances used can be:

- **Toxic**
- **Corrosive**
- **Irritants**
- **Harmful**

Ensure you are aware of the correct precautions to take when working with hazardous liquids or substances.

Control of Substances Hazardous to Health (COSHH) safety data sheets are available for all of these substances stating the precautions and actions to take.

Never put your fingers near the fan, even when the engine is stopped. The fan might start automatically even when the engine has been turned off.

ENGINE COOLING

For an engine to work it must burn a mixture of fuel and air (combustion). This generates a great deal of heat and can momentarily rise as high as 1800°C which is higher than the melting point of aluminium (660°C). To prevent this, the engine must be cooled to an appropriate operating temperature.

Name the two main types of cooling systems:

- _____ is found mainly on motorcycles, petrol driven mowers, strimmers and some older vehicles.

- _____ is used for cars and commercial vehicles.

Heat flows from a hotter place to a cooler place. An example of this is a saucepan of cold water on a hot hob. The heat from the hob will warm the saucepan. The heat from the saucepan will then pass to the water making it boil.

Heat can be transferred in three ways and in a vehicle's cooling system all three can occur at the same time.

Give examples of each one:

1 Conduction: _____

_____

2 Convection: _____

_____

3 Radiation: _____

_____

LIQUID COOLING SYSTEMS

Identify and name the main components of the cooling system.

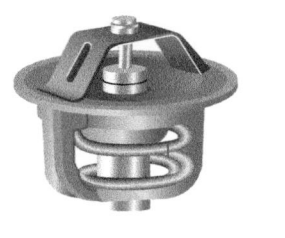

_____

_____

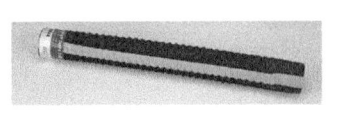

_____

_____

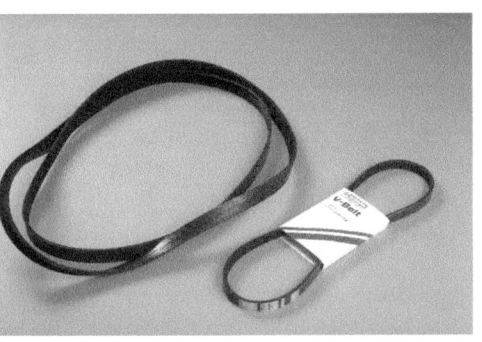

_____

_____

VEHICLE COOLING SYSTEM COMPONENTS

On a workshop vehicle, locate the components from the previous page and draw a basic diagram of the vehicle's cooling system. Indicate which direction the coolant flows. Explain the main function of each of the components shown in your diagram.

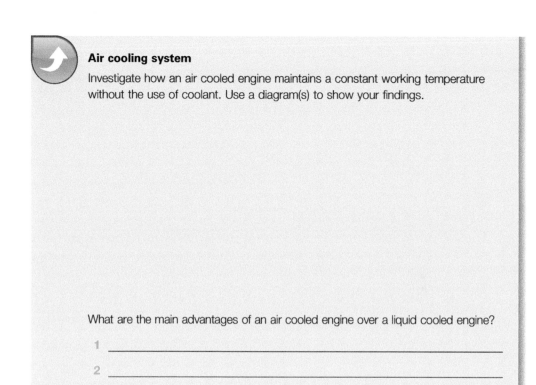

Air cooling system

Investigate how an air cooled engine maintains a constant working temperature without the use of coolant. Use a diagram(s) to show your findings.

What are the main advantages of an air cooled engine over a liquid cooled engine?

1 _____

2 _____

3 _____

4 _____

THERMOSTAT

This is a valve that is temperature sensitive. It controls the flow of coolant to the radiator. State THREE important reasons why it is fitted:

1 _____

2 _____

3 _____

Wax element type thermostat

This type of **thermostat** uses a special type of wax that is contained in a strong metal cylinder into which passes the reaction or fixed pin. It is surrounded by a rubber sleeve which acts as a seal at the upper end of the reaction or fixed pin.

Name the main parts on the diagram of a wax element type thermostat below.

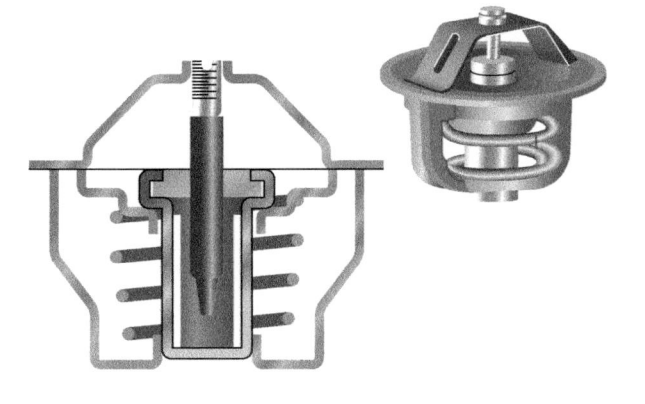

Indicate the flow of the coolant through the thermostat.

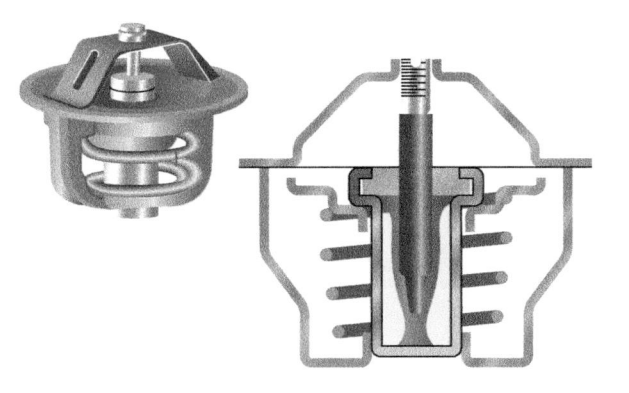

Some thermostats have a small hole and pin in the valve disc. What is the purpose of this?

_____

_____

Complete the following paragraph:

All substances exist in one of three states: solid, liquid or gas. In a _____ the wax changes from a _____ to a _____ when heated.

Describe the sequence of events that occur in the engine to cause a change of state in the wax type thermostat.

_____

_____

_____

_____

Why should you take care when removing a cooling system component?

_____

When removing a radiator, which part can easily become damaged?

_____

Before removing the components, what should the coolant be drained into?

_____

Where would you find technical information relevant to the vehicle?

_____

_____

Thermostat operation test: Removing, testing and replacing a thermostat

Check the serviceability of a thermostat when it has been removed from the vehicle.

Thermostat make _____ Type _____

Specified opening temperature _____

Actual temperatures: opening _____ closing _____

Visual defects (if any) _____

Serviceability _____

When replacing a thermostat, check that the replacement has the same temperature rating as the original. Before refitting any new components always clean all surfaces and replace any gaskets or 'O' rings.

PRESSURE CAP

The pressure cap is fitted to the radiator, header tank or expansion bottle and seals the system. As the coolant heats up, the pressure increases. The cap is designed to retain a certain amount of pressure.

When topping up a cooling system the engine should be stopped. **Do not remove the pressure cap when the engine is hot** as the temperature may be over 100°C and any steam given off will be even hotter.

Even if the engine is only warm, keep your hands and clothing well away from the fan or any moving parts.

Do not allow antifreeze to come into contact with paintwork as it will damage it.

What effect will increasing the pressure in a cooling system have on the boiling point of the coolant?

_____

_____

What could happen when the coolant temperature and pressure begin to reduce?

_____

_____

Label the diagram shown using words from the list below and indicate the flow of coolant as the pressure cap operates.

Pressure release spring	Vacuum release valve	Diaphragm
Pressure release valve	Pressure seal	locking spring disc

Explain why some vehicles have an expansion bottle and some do not.

_____

_____

_____

_____

Remove and refit radiator

In small groups or in pairs, place the headings listed below in the correct order to describe the procedure for removing and refitting a vehicle's radiator.

Briefly state how to carry out each of the following operations taking into consideration health and safety and reducing the risk of any damage. State any tools required.

Inspect the radiator   **Remove the radiator**   **Drain the coolant**
Replace the radiator   **Refill the system**

Procedure	Description of operation

Procedure	Description of operation

RADIATOR CONSTRUCTION

Radiators are usually mounted where the path of greatest airflow can be found. As the air passes over the fins of the radiator core, heat is carried away, cooling the liquid before it returns to absorb more heat from the engine and components.

Identify the following components on the diagram below.

O-ring gasket   **Inlet tank**   **Radiator core**
Bending tangs   **Drain cock**
Outlet tank   **Transmission oil cooler**

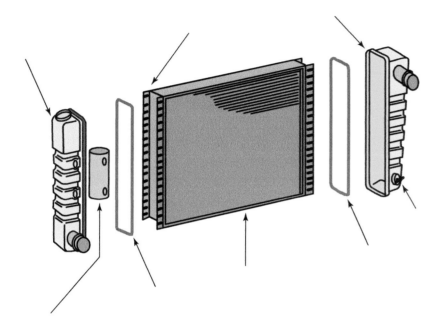

Name the materials that are used in the construction of a radiator.

_____

Name the type of radiator shown above. _____

Research and draw the other most popular design of radiator. What is this called? _____

Indicate on the diagrams the direction of flow of the coolant. _____

Complete the following paragraphs on radiator components using the word bank below (note: there are two distractor words):

water pump	tubes	vertical	cooled
radiator	core	air	tubes
tanks	fin	increases	
horizontal	surface area	coolant	

The construction of the radiator _____ consists of a number of _____ that carry

coolant between the _____. The tubes can be either _____ or _____.

The small thin fins are attached to the _____. The number of fins _____ the

_____ exposed to _____.

The _____ that emerges from the bottom of the radiator has been _____. It travels

through the bottom radiator hose to the _____ inlet, then around the engine again.

Pressure testing a cooling system

A pressure test is carried out to identify any external or internal leaks. Explain how to conduct a pressure test on a liquid cooling system.

● _____
  _____
  _____

● _____
  _____
  _____

● _____
  _____
  _____

83

COOLING FANS

Most cooling systems (water or air) are fitted with a fan to help generate airflow through the radiator.

What can be used to drive the cooling fans?

_____

_____

RMP For the removal and replacement of an auxiliary drive belt it is always good practice to consult the manufacturer's recommended procedure.

Why is the permanently driven type of fan rarely used on modern vehicles?

_____

_____

When might a fan be essential to provide extra cooling?

_____

_____

Why would a vehicle have more than one fan fitted?

_____

Apart from the cooling fan what else can the drive belt be used for?

_____

_____

TIP Fan blades can be rigid or flexible. Rigid blades tend to be noisy and use more energy. This noise can be reduced by using irregular spacing of the fan blades.

What is fitted to some vehicles to help direct the air flow over the cooling fins?

_____

_____

In groups, research the two main ways that vehicle manufacturers control the fan so that it is not always working at full speed and wasting energy.

1 _____

_____

2 _____

_____

_____

WATER PUMPS

Water pumps, or as they should correctly be called, 'impellers', are usually bolted to the front of the cylinder block. They can be driven from the crankshaft by the timing belt, auxiliary belt or fan belt.

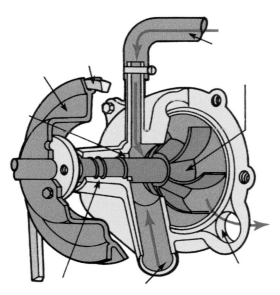

Name the different parts of the water pump shown opposite.

fan belt	**bypass**	**inlet from radiator**
seal	**impeller**	**outlet to cylinder block**
bearings	**drive pulley**	

Complete the following paragraphs using words from the following word bank (note: there are two extra distracter words):

local	passages	engine
drive belt	valves and ports	water jackets
impeller	coolant	
blades	circulate	

The rotor or _____ has fan-like blades. The rotor is turned by the _____. This rotation forces the coolant to _____ around the engine. It is pushed through the outlet _____ into the engine and the passages called _____.

Water jackets are passages designed into the engine block and cylinder head that surround the cylinders, _____ and direct the _____ to all of the required parts of the engine that need to be cooled and stop _____ overheating.

 TIP Many hybrid vehicles use an electrically operated water pump.

 In small groups discuss why you think electrically operated water pumps are used on hybrid vehicles. You will need to think about how a hybrid vehicle operates and how it is designed to be economical in and around a town or city.

ANTIFREEZE

Antifreeze is a mixture of ethylene glycol together with some protective additives. When added to water, ethylene glycol is able to lower the freezing point of water, as well as raise its boiling point. The additives that are added to the mixture play a very important role within the engine.

TIP Not all antifreezes are the same. Refer to manufacturers' specification before refilling the cooling system.

Antifreeze can be toxic if swallowed. It can damage your internal organs.

From the word search below identify these terms.

```
C C N T I E N Z A I T O C O
I E E R L B O I L I N G R T
O A T O G Z I I U I N C C A
B B N I N R T I O I I C O N
L O C I I O C I Z N R F C D
U F A I M B A E E N A O N I
G L I M A T E E O I C U M G
N O I S O R R O C I T N A I
N I C O F A C T T N O O L M
I N I A I N I O R B R I C R
O G T O N C D O U E R R N O
C A C C S R I S R G C F I U
M F N I T C C I R A Z B I I
L U B R I C A N T I N F O C
```

LUBRICANT
ANTICORROSION
ACIDICREACTION
BOILING
FREEZING
FOAMING

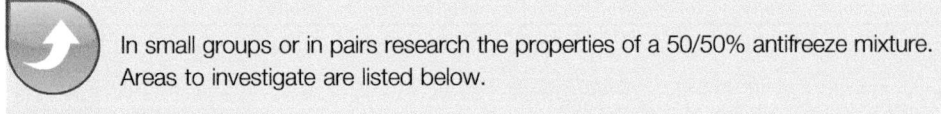

In small groups or in pairs research the properties of a 50/50% antifreeze mixture. Areas to investigate are listed below.

Determine the:

● Freezing point: _____

● Boiling point: _____

What would happen if the mixture is too strong or too weak?

_____

_____

_____

Why should the coolant be changed at recommended intervals?

_____

EFFECTS OF FREEZING

When water freezes it expands with such force that it can crack a cylinder block and head causing extensive damage. It could well mean a new engine!

 If antifreeze is spilt on paintwork, wash it off immediately. Antifreeze will strip paint.

TIP Antifreeze is a carefully balanced mixture of ethylene glycol with rust inhibitors, corrosion inhibitors, scale inhibitors, pH buffers for the acid to alkali balance, anti-foaming agents, and reserve alkalinity additives.

ANTIFREEZE TESTERS

Below are two main types of equipment used to test the specific gravity of antifreeze.

Hydrometer

Refractometer

Inspection and testing

Place the statements below in the correct order.

1 Check these readings with a reference card.
2 How much antifreeze do you need to **top up** to a safe percentage?
3 Add antifreeze to give the correct percentage and check again.
4 With an **antifreeze tester,** take some coolant from the system.
5 What percentage of antifreeze is there in the system?
6 Run the engine to its normal running temperature.
7 Note the readings of the antifreeze tester.

1 _____

2 _____

3 _____

4 _____

5 _____

6 _____

7 _____

Check the proportion of antifreeze in a selection of different vehicles' cooling systems and complete the table below.

Vehicle make/model	First mixture % reading	% antifreeze needed to be added	Final reading

HOSES

Most hoses are made of rubber reinforced with a layer of fabric to withstand the operating pressures and changes in temperatures of the components. They can be moulded to special shapes and are designed to be flexible to allow for the relative movement of components such as the engine and radiator.

Sketch the construction of a flexible hose.

 **RMP** During a routine vehicle maintenance service, visually check all coolant hoses for leakage, splits, swelling and chafing.

 **RMP** Check hoses for burned or chafed areas. Always feel under the hoses. Squeeze the hose to find any cracks and breaks that might not be easily seen.

Identify the common faults shown on the hose diagrams below.

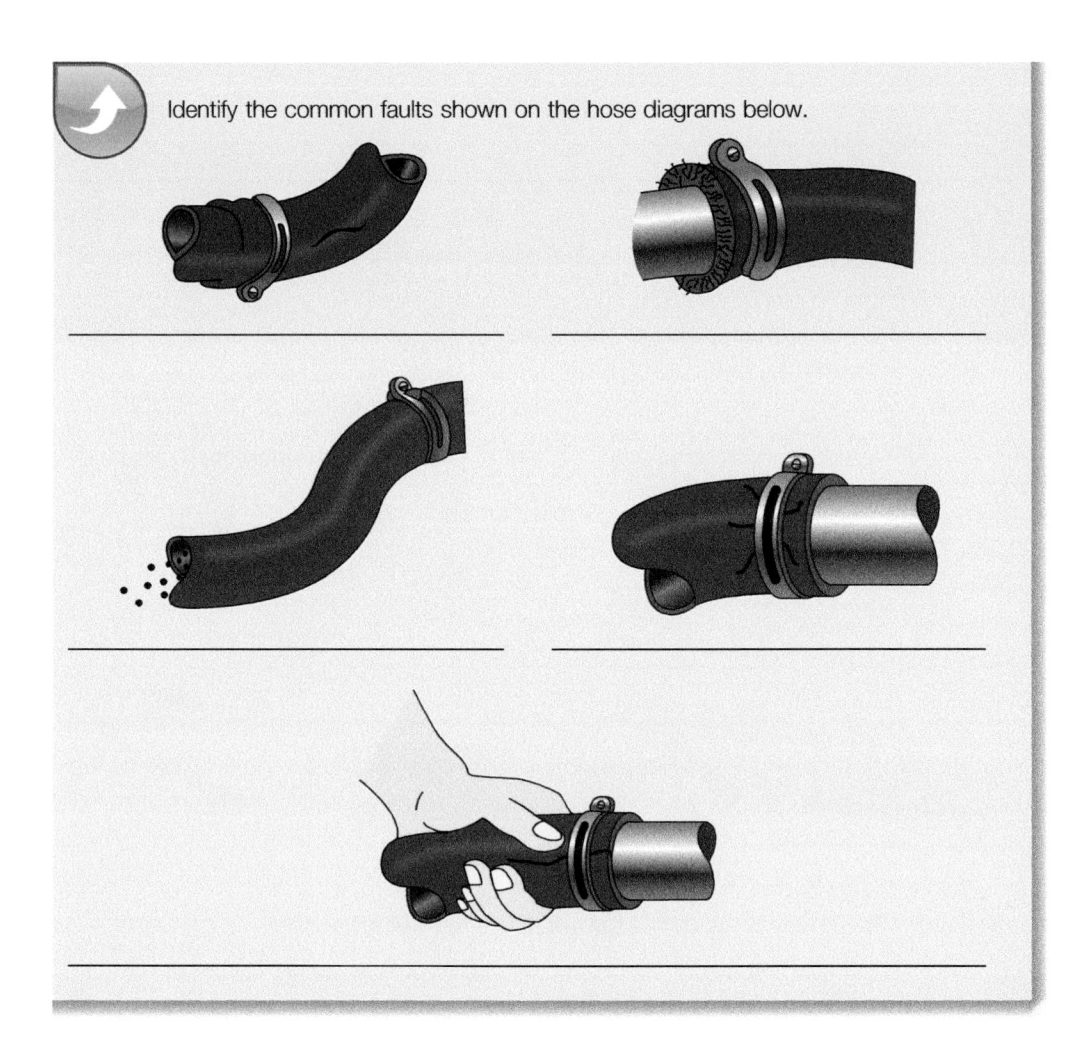

If a hose feels 'soft' this could mean that:

_____

_____

_____

_____

If a hose feels 'hard' this could mean that:

_____

_____

If a hose feels 'swollen' or 'oil-soaked' this could mean that:

_____

_____

_____

Wire-type clamp

Squeeze hose

Multiple choice questions

Choose the correct answer from a), b) or c) and place a tick [✓] after your answer.

1 Which component is used to raise the boiling point of the coolant?

a) Thermostat [ ]

b) Water pump [ ]

c) Pressure cap [ ]

2 What percentage of antifreeze is added to a cooling system?

a) 25% [ ]

b) 40% [ ]

c) 50% [ ]

3 Which of the two systems of cooling an engine will allow it to heat up the quickest?

a) Air [ ]

b) Liquid [ ]

c) Gel [ ]

4 What can be used to increase the amount of air passing over a radiators' fins?

a) Cooling fan [ ]

b) Heating fan [ ]

c) Impeller fan [ ]

5 What would happen to the engine if the thermostat was stuck in a closed position?

a) The engine would operate at normal temperature [ ]

b) The engine would operate at below normal temperature [ ]

c) The engine would operate at above normal temperature and overheating could occur [ ]

PART 2: LUBRICATION

Key terms

Wet sump or reservoir Oil is returned from the engine by gravity and collected in a sump below the engine.

Dry sump Oil is supplied and returned to a separate oil tank away from the engine.

Full-fluid film A film of lubricant that is sufficiently 'thick' so that no metal-to-metal contact takes place.

Hydrodynamic lubrication Using the natural movement of the oil 'wedge' to separate the surfaces of highly loaded bearings when shafts rotate.

Boundary A term applied where the film of lubricant is applied by splash and mist with the possibility of some metal-to-metal contact.

Viscosity The resistance to flow or 'thickness' of a liquid. It can also be described as its resistance to shear.

Multigrade Oil which meets the viscosity requirements of several different single-grade oils.

Viscosity index A number which indicates how the viscosity of a liquid changes with temperature.

```
W E U L N I I U D O C B S D
M E O M S A I N R O E A O C
X E D N I Y T I S O C S I V
R E D U R V O E U U N M I Y
M A R A E V Y C M R A U N N
Y R E W R C H R G N T L T C
A M W E D G E U Y A S T M S
E E S T W E Y D R D I I R I
S E P A S E O R A E S G G D
R U L G R R T Y D T E R U E
E R A A D M T S N S R A S T
G E S Y A N S U U T X D W I
O S H D F I L M O M D E R C
Y L D R M U P P B H P V W Y
```

WETSUMP
VISCOSITYINDEX
GRADE
MULTIGRADE
RESISTANCE
MIST
SPLASH
FILM
BOUNDARY
WEDGE
HYDRODYNAMIC
DRYSUMP
SUMP
RESERVOIR

The need for lubrication

A film of oil between moving parts reduces metal-to-metal contact which causes friction, heat and wear. It also cools the metal surfaces that it comes into contact with. In engines, oil also carries away tiny metal and carbon particles.

Why does the oil carry away contamination?

_____

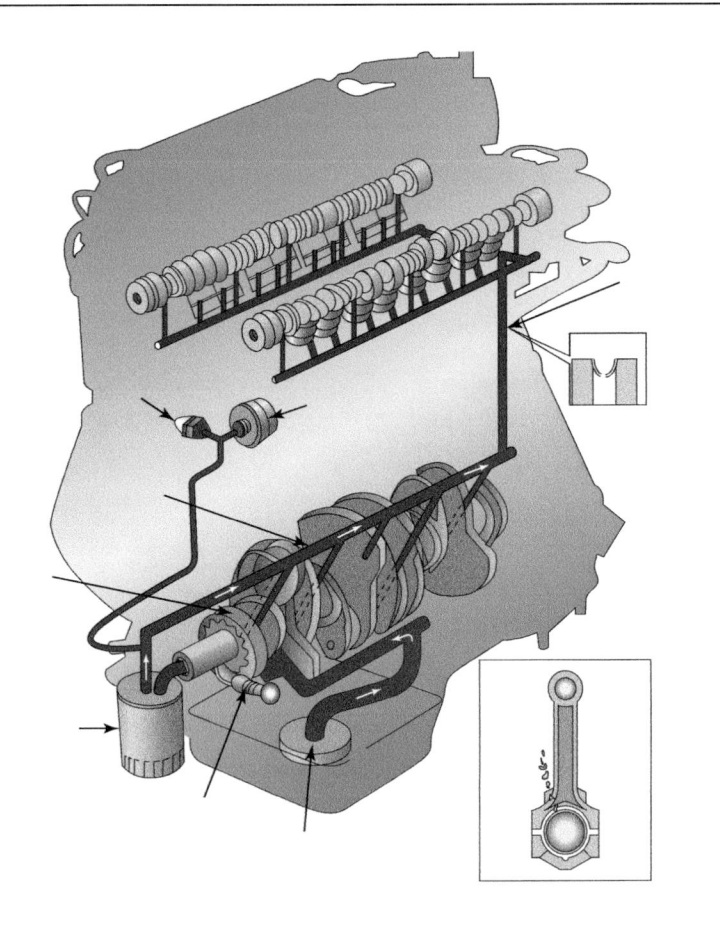

On the picture of an engine lubrication system shown opposite indicate the direction the oil will travel around the engine. Then label the components listed below.

main oil gallery	**primary oil strainer**	**oil pressure restrictor**
oil pressure switch	**main oil filter**	**oil pressure transducer**
oil pump	**pressure relief valve**	

Listed below are the main components of a wet sump lubrication system. In small groups research what each of them does and how they work.

Sump/reservoir

_____

Pick-up pipe

_____

_____

Oil pump

_____

_____

_____

Oil pressure relief valve

_____

_____

Oil filters

_____

Bypass valve

_____

_____

Galleries

_____

_____

Oil cooler

_____

_____

_____

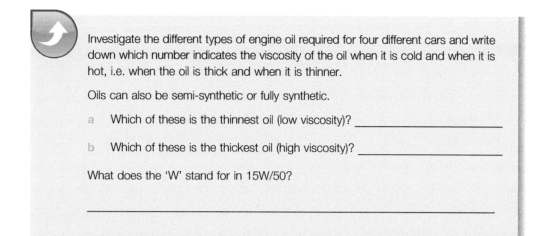

Investigate the different types of engine oil required for four different cars and write down which number indicates the viscosity of the oil when it is cold and when it is hot, i.e. when the oil is thick and when it is thinner.

Oils can also be semi-synthetic or fully synthetic.

a   Which of these is the thinnest oil (low viscosity)? _____

b   Which of these is the thickest oil (high viscosity)? _____

What does the 'W' stand for in 15W/50?

_____

_____

_____

TYPES OF OIL

Oils used in vehicles are mostly based on mineral or natural oils. They are graded by their viscosity and their quality. **Viscosity** is a measure of how easily a liquid flows or how 'thick' it is. It can also be described as its resistance to shear. The _thinner_ the oil, the _lower_ the viscosity number.

It is vital to use the correct type of oil.

ENGINE OILS

Multigrade oils

Modern oils are blended with additives, called **viscosity index** improvers, to form **multigrade**, or multi-viscosity oils. They provide better lubrication over a wider range of climatic conditions than basic mineral oils.

All oils become 'thinner' as they are heated. Multigrade oils contain additives so that they 'thin' more slowly.

Name one advantage of using a thinner oil in an engine?

_____

Synthetic and semi-synthetic oils

Synthetic oils are often used in the modern vehicle as they offer better protection against engine wear and can operate at much higher temperatures. They have better low temperature viscosity, are chemically more stable and allow for closer tolerances in engine components without loss of lubrication.

What is the main difference between synthetic oils and multigrade oils?

_____

_____

Synthetic oils also last considerably longer, extending oil change intervals up to 10 000 miles or more.

Name one advantage and one disadvantage of extended oil changes.

Advantage: _____

Disadvantage: _____

 How is synthetic motor oil made? Visit **www.ehow.com** to find out.

Oils are graded or classified by the Society of Automotive Engineers (SAE), or classified by the American Petroleum Institute or API service classification. Oils for spark-ignition engines carry a prefix S, and diesel or compression-ignition engines use C. These organizations ensure that the oil is of the required standard.

 Manufacturers give different specifications for oils, so it is vitally important that new oil is checked carefully.

Complete the following word search:

```
S P E E G R M E E I R S P U C U B T R
P H I D Y G R P C L S N R R A A A D U
E V L A V F E I L E R E R U S S E R P
S E V R Y L T P U P C E R S I E U R T
V T R G Y E L P U I L P T P P S P S L
I C N I H S I U V O U A G T K T A S G
S O L T C G F K O E D P U S I P I S O
C L N L A M U C I O I B W I E R E U I
O Y P U D S R I A L E Y E I S E P S E
S L Z M S U M P E Y O U C R I C E E T
I E D W U I P P T P R V U S V L V R R
T R I S P Y S E R C L I A E P E A C E
Y E P R T E U I H V A G A O E Y U I I
K E W L E U T B U R E E O I W L S F T
Y R Y L S L K Y E G E P M C K R R P A
H D U N M T L P G F S R P I N L L U S
Y U P T E E M A I I L T A E R P B B I
E S S R U A S G M U C Z Y Y P U G T
U P L A P R P S T R P I C U P P I E V
```

SYNTHETIC
BYPASS
VISCOSITY
PRESSURE RELIEF VALVE
DRY
WET
MULTIGRADE
GALLERIES
GAUZE
FILTER
SUMP
PICKUPPIPE
COOLER
PUMP

Lubrication is based on maintaining a fluid film between components whenever possible.

The two drawings below each show a magnified view of part of a bearing face. Complete the drawings and indicate above each one to show what is meant by 'full-fluid film' and 'boundary' lubrication.

Full-fluid film lubrication   Boundary lubrication

Boundary lubrication is where the layer of oil is sometimes only one molecule thick and could easily break down.

What could occur if the film of oil was broken down?

_____

_____

 Each system on each vehicle needs a specific amount of the right fluid. This is the refill capacity. It must be strictly adhered to as damage to the engine can happen if it is over-full or under-filled.

ENGINE OIL MAINTENANCE

Changing the engine oil and the filter

Oil filters are usually replaceable canisters. They **must** be replaced when the engine oil is changed. The change period differs according to manufacturers' specifications.

It is easier to do this when the vehicle is raised on a ramp or lift.

Before you start, name the equipment you will require.

1 _____

2 _____

3 _____

4 _____

5 _____

6 _____

7 _____

 Changing the oil is best undertaken when the engine _is warm_. If it is _cold_, the oil will not flow easily. If it is very _hot_, the oil could scald you. Ensure the vehicle is secure and the ignition keys removed so the engine cannot be started accidentally.

Oil change procedure

Ten headings describing the main steps for this procedure are listed below. Before you undertake this task, place the headings in the correct order

1 **Fill up with new engine oil, to the correct level**
2 **Smear the rubber sealing ring of the oil filter with new engine oil**
3 **Replace the oil filter**
4 **Drain the engine oil; replace and tighten sump plug**
5 **Remove the filter**
6 **Check the oil level using the dipstick. Top up, if necessary**
7 **Position the drain tray**
8 **Start the engine, watch if the oil light goes out and check for leaks**
9 **Clean the vehicle of any spilt oil**
10 **Dispose of waste material properly**

1 _____

2 _____

3 _____

4 _____

5 _____

6 _____

7 _____

8 _____

9 _____

10 _____

 By law, you must not pour old oil into drains, and you must not place oil filters in ordinary waste bins.

How should oils and oil filters be disposed of?

_____

_____

OTHER LUBRICANTS AND FLUIDS

Many systems in a car need **lubricants** or **fluids.** A common task for a technician will be to look under the bonnet and check the **levels** of these lubricants and fluids. On modern vehicles, items that need regular checks often have colour-coded tops.

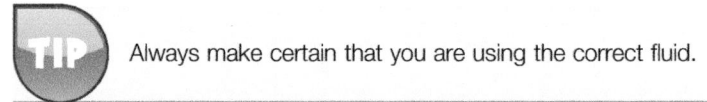 Always make certain that you are using the correct fluid.

If, for example, you put screen washer fluid in the cooling system the engine will freeze or over-heat.

Apart from the engine, look at all other vehicle components and systems and list the different types of oils that are used. Technical information may need to be found to help you, i.e. a workshop manual if necessary.

State typical lubricants and their grades found in the following components/systems.

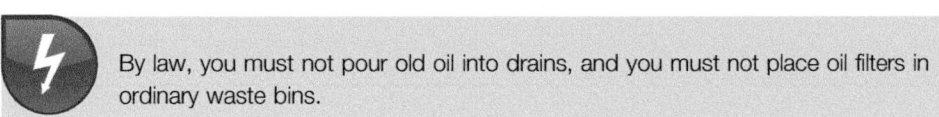

Component or system	Oil type and specification
Engine	_____
Manual gearbox	_____
Automatic transmission and power steering	_____
Cooling system	_____
Braking system	_____

Multiple choice questions

Choose the correct answer from a), b) or c) and place a tick [✓] after your answer.

1 **Which component provides the first or primary oil filtration?**

 a) Oil filter [ ]

 b) Engine breather [ ]

 c) Oil strainer [ ]

2 **Oil thickness is classified by its:**

 a) Colour [ ]

 b) Viscosity [ ]

 c) Velocity [ ]

3 **Refilling an engine with the incorrect type of oil will cause it to:**

 a) Seize [ ]

 b) Tick over faster [ ]

 c) Extend the oil change intervals [ ]

4 **Before commencing work on a vehicle, VPE should be fitted. What does this stand for?**

 a) Personal protective equipment [ ]

 b) Vehicle protective equipment [ ]

 c) Visual protective equipment [ ]

5 **What does 'W' stand for in 5W/50?**

 a) Weather [ ]

 b) Warm [ ]

 c) Winter [ ]

6 **What type of engine oil is normally used in a modern vehicle engine?**

 a) Multi-type [ ]

 b) Multigrade [ ]

 c) Multi-temp [ ]

SECTION 3

Exhaust systems

USE THIS SPACE FOR LEARNER NOTES

_____

_____

_____

_____

_____

_____

_____

Learning objectives

After studying this section you should be able to:

● Identify exhaust system components used on a modern spark ignition vehicle.

● State the purpose of the major components of the exhaust system.

● Describe the correct procedure and sequence for the removal and refitting of vehicle exhaust systems and components.

Key terms

Catalytic converter   Converts harmful gases into less harmful gases.
Silencer   Reduces engine noise to an acceptable level.
Carbon monoxide   A pollutant gas that can kill following long exposure.
Lambda sensor   Measures the oxygen content in the exhaust gases.

www.bosal.uk

www.eurocats.co.uk

```
A N E N Y L P F T C E H
T O M O U N T I B B T T
D I C I T Y L A T A C C
I S C T B L A M B D A P
S N O B R A C O R D Y H
I A N R A E S P R L A S
L P V O C T I V G O I S
E X E S K P E A D F A S
N E R B E S S K T I B L
C C T A T E G A S N R B
E A E T S E L F F A B C
R V R U S T E D O M G C
```

CATALYTIC   SILENCER   PIPE
CONVERTER   BAFFLES   GASKET
LAMBDA   MOUNT   BRACKETS
HYDROCARBONS   EXPANSION   ABSORBTION
MANIFOLD   GASES   RUSTED

THE EXHAUST SYSTEM

The purpose of the exhaust system is to quieten or silence the noise made by exhaust gases leaving the engine. The exhaust system is also designed to direct the exhaust gases and heat away from the passenger compartment of the vehicle, normally leaving at the rear of the vehicle. It also helps to improve the performance of the engine and to improve fuel consumption.

The exhaust consists of a series of pipes that link the engine to a silencer and a catalytic converter.

When running an engine in the workshop, always use an exhaust gas extraction system.

Correctly label the diagram of the exhaust system with the following component names:

expansion box   Heat shield   Gasket
flexible pipe   Mounting brackets   catalytic converter
absorption box   Manifold mounting flange

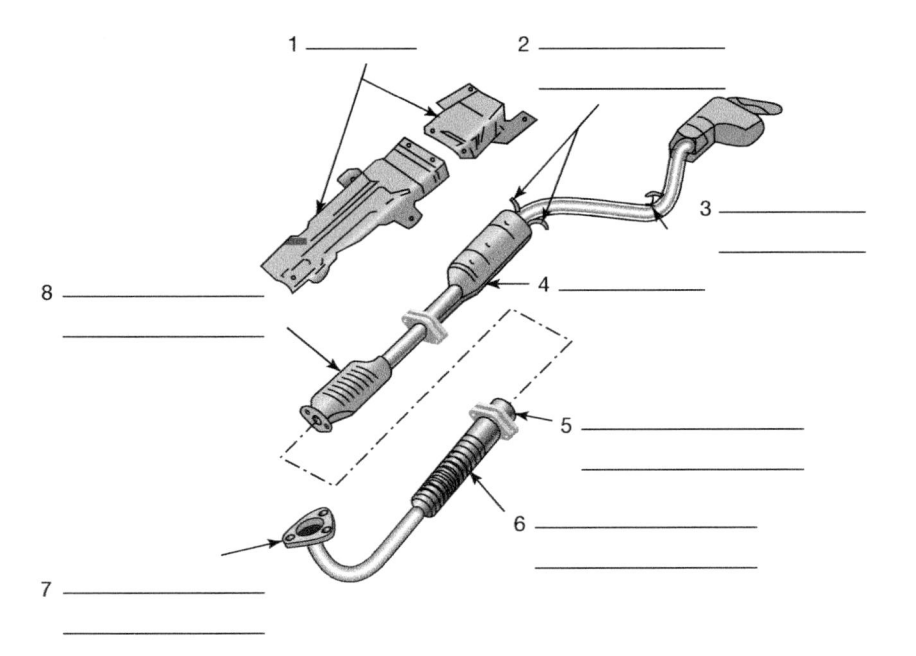

Exhaust systems can corrode from both the inside and the outside. The life of the exhaust system depends on how often and how far the vehicle is driven, rather than how long the system has been fitted. If a vehicle is used for short trips, especially the stop–start variety, more than for long distance journeys, the exhaust will tend to corrode in a shorter time. This is because moisture, or water vapour, is present in the exhaust and when you shut down your engine whatever water vapour is in the pipes condenses and turns back into a liquid. On a short trip the water is not able to get hot enough to turn back into water vapour so it stays in the system and rusts the exhaust pipes.

Each of the exhaust system components has a specific function. On the following chart draw lines to match the component to its function:

Silencer box	This is the final part of the exhaust and is responsible for directing the gases away from the vehicle, passengers and passing pedestrians.
Exhaust system brackets	Changes three harmful gases into harmless gases.
Catalytic converter	Mounts the exhaust system; allowing for flexible movement when the engine is running.
Tail pipe	These are designed to reduce engine noise to an acceptable level.

Manifolds

These connect the exhaust ports to the downpipe and are made from cast iron to give them strength and the ability to endure the very high temperatures of the exhaust gases as they leave the engine.

The manifold helps to reduce combustion noise as well as transferring the gases and heat away from the engine to the exhaust system.

An exhaust manifold

Silencers

Complete and correctly label the following cross-sectional silencer diagrams:

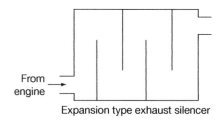

From engine →

Expansion type exhaust silencer

From engine →

Absorption type exhaust silencer

Cross-section exhaust diagrams

RMP Visually check the exhaust system for security and condition. With the engine running check for any leaking gaskets, joints or leaks from corroded components. Physically tap the silencers to check the baffles are secure.

Leaking gaskets and seals are often found between the exhaust manifold and pipe and lead to failure of the MOT emissions test as the exhaust gas machine will detect a leak.

Catalytic converter

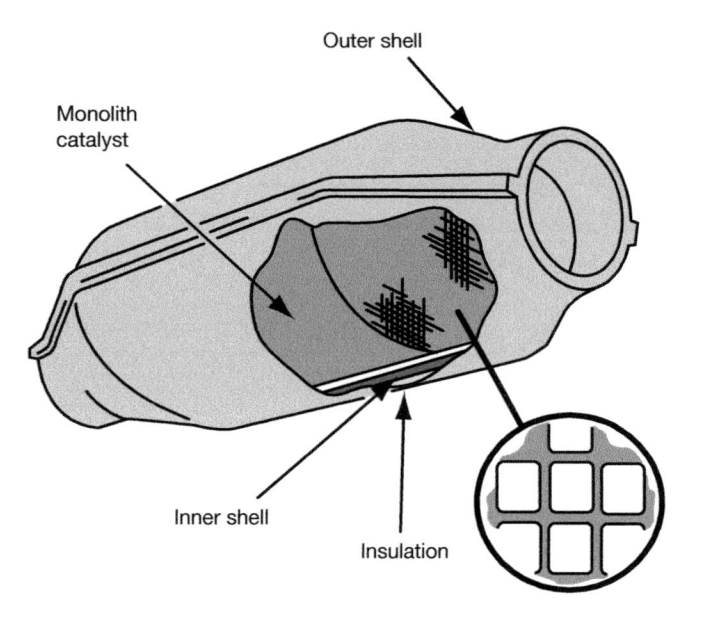

Outer shell

Monolith catalyst

Inner shell

Insulation

The main function of a catalytic converter is to convert the dangerous exhaust gases into non-toxic gases. Most manufacturers use a three-way catalyst (TWC). Externally the catalytic converter looks like a normal silencer box. Internally the 'cat' (as it is generally known), has a honeycomb effect which converts the three pollutant gases into harmless ones. These are:

Carbon monoxide to _____ (CO to CO_2)

Hydrocarbons to _____ (HC to H_2O)

Oxides of nitrogen to _____ (NO_x to N_2)

Always fit Original Equipment (OE) catalytic converters. If a non-genuine one is fitted, problems can be experienced with the engine management and emission control systems.

Lambda sensor

A lambda sensor is also called an oxygen sensor

What is the purpose of this sensor?

_____

There are normally two of these fitted to a modern spark-ignition vehicle. Where would they be

found? _____

REPLACING EXHAUST SYSTEMS

Insert the missing words into the following paragraphs using the word bank below (note: there are two extra distracter words):

management	exhaust	seized	components
nuts	catalytic	vehicles	replace
exactly	broken	original	specialist
penetrating	replacing	roughly	pipes

When replacing exhaust _____ make sure they _____ match the _____

parts. This is especially important for the _____ converter, if it is not correct it can cause

engine _____ running problems.

When removing _____ systems from _____ it is common to come across rusted

parts. These can include _____ and exhaust _____ may be rusted together.

_____ exhaust system components may require the use of _____ tools.

Before attempting to _____ rusted and _____ nuts and bolts, spray them with a _____ oil.

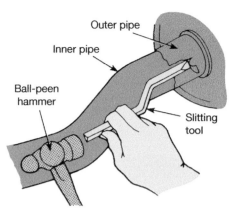

Outer pipe

Inner pipe

Ball-peen hammer

Slitting tool

Using a splitting tool on a rusted silencer

Identify the specialist tools used to replace an exhaust system.

_____   _____

_____   _____

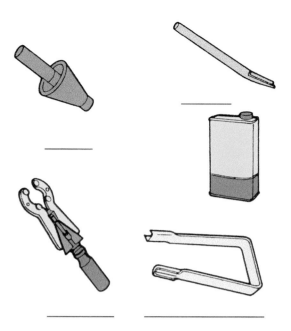

_____

_____

_____   _____

The correct vehicle protective equipment must be used on the vehicle. This will include:

_____

REMOVAL AND DISMANTLING OF THE EXHAUST SYSTEM

It is important that an exhaust system is correctly supported and removed in a set sequence. This reduces the time, effort and likelihood of injury. List a suitable procedure for removing and dismantling an exhaust system:

● _____

● _____

- _____
- _____
- _____
- _____
- _____

Wear goggles and thick leather gloves when working on exhaust systems. These will prevent rust from getting into the eyes and will stop hands from becoming severely burnt.

Multiple choice questions

Choose the correct answer from a), b) or c) and place a tick [✓] after your answer.

1 **The catalytic converter converts:**

a) Three harmful gases to harmless gases [ ]

b) One harmful gas to a harmless gas [ ]

c) Three harmless gases to harmful gases [ ]

2 **Which component monitors the oxygen content of the exhaust gases?**

a) Expansion box [ ]

b) Lambda sensor [ ]

c) Absorption silencer [ ]

3 **Which ONE of the following components transfers heat and gases from the engine to the exhaust system?**

a) Catalytic converter [ ]

b) Manifold [ ]

c) Tailpipe [ ]

REFITTING THE EXHAUST SYSTEM

Insert the missing words in to the following paragraph using words from the word bank below (note: there are two extra distracter words):

excessive	gaskets	miles	components	mountings
rear	smell	whole	customer	
positioning	grease	checked	sealing	

Replace all _____ with new ones. Use exhaust _____ compound on pipe joints. The

exhaust must be _____ for correct positioning and location, be sure its _____ is clear

of bodywork and other vehicle _____. The engine needs to be run and the _____

exhaust system checked for leaks and _____ noise. Check that there are no _____

marks on the vehicle before handing over to the _____. Inform the customer that there may

be an unusual _____ for a few _____ whilst the new exhaust system settles in.

SECTION 4

Spark ignition systems maintenance

USE THIS SPACE FOR LEARNER NOTES

_____

_____

_____

_____

Learning objectives

After studying this section you should be able to:

- **Understand how to work safely when working on vehicle ignition systems.**
- **Understand vehicle ignition systems.**
- **Undertake the replacement of vehicle ignition components.**

Key terms

Low tension circuit A circuit that carries normal battery voltage.
High tension circuit A circuit that carries anything up to 40 000 volts.
High tension leads Carry high voltage from the distributor to the spark plugs.
Dwell The period of time when electricity is passed into the coil (typically 3–6 milliseconds).
Coil An ignition coil is able to transform battery voltage to the very high voltages required for a spark to cross the electrode gap of a spark plug.
Amplifier A device used to increase the electrical signal in an electronic ignition system.
Transistor A semi-conductor which can be used to switch electronic circuits and also amplify voltage.
DIS Distributor less ignition system.
Distributor A mechanically driven component, which distributes high voltage to the spark plugs.
Contact breaker A mechanical switching device used in older ignition systems.
Air gap The gap between the electrodes of a spark plug or reluctor and pick-up.

Remember – when working with ignition systems you may be dealing with:

- Very hot surfaces
- High voltages that can cause serious injury, not only from electrical shock (up to 40000 V), but when you jump your head could hit the bonnet or your fingers could get caught in other moving parts
- Components which are carcinogenic or toxic
- Rotating components
- Waste disposal

TIP   Ignition systems can have the components located in different places so using specific vehicle technical data can help to identify them and their location.

IGNITION SYSTEM

An ignition system is designed to provide a very high voltage at the spark plugs to ignite the air/fuel in the combustion chambers. The spark must be supplied at the right time, and it must have enough energy to ignite the mixture.

Fill in the gaps in the following paragraph:

The system has two circuits. The primary or low tension circuit where the power comes from

the _____ and _____, and the secondary or high tension circuit which produces the

_____ voltage for the spark _____ via the _____.

There are three main types of ignition systems:

1 Older vehicles use mechanical contact breaker points. Where are these located?

_____

2 Modern vehicles use either electronic ignition which triggers an electronic transistor to create a spark when a signal is received from the distributor; or

3 Direct ignition which has no distributor and uses.

To deliver high voltages, what is required by the spark plugs?

_____

The drawing below is of a basic ignition system. Connect all of the components together and use arrows to indicate the direction that the electricity travels in.

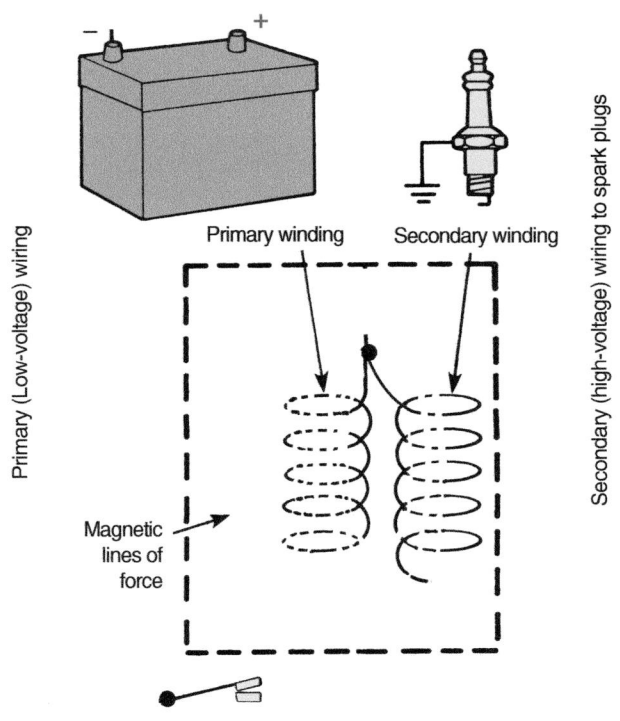

Identify ignition system components in the above diagram.

From the pictures shown name the ignition components.

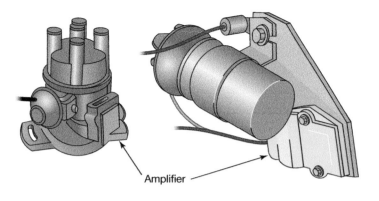

Amplifier

_____ _____

_____

_____

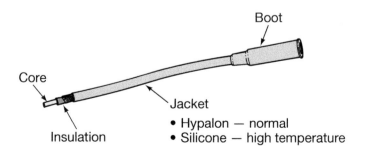

Boot

Core

Jacket

Insulation

• Hypalon — normal
• Silicone — high temperature

_____

_____

All ignition systems consist of two interconnected electrical circuits, low voltage (LT) and high voltage (HT).

Indicate from the components named below whether they are connected to the HT or LT ignition components.

Low voltage (LT)	Ignition system components	High voltage (HT)
	Ignition switch	
	Spark plugs	
	High voltage cables	
	Ignition coil primary winding	
	Triggering device and control	
	Battery	
	Distributor cap and rotor	
	Ignition coil secondary	
	Winding	

Ignition timing

The electrical system in your car supplies voltage to the **coil** and ignition unit. The distributor will trigger a spark for each cylinder, via either a mechanical trigger (points), or via an electronic module at the correct time.

When is the spark generated?

_____

_____

What carries the high voltage from the distributor cap to the relevant spark plug?

_____

In modern ignition systems the electronic control unit (ECU) monitors sensors all over the engine to ensure the engine is operating efficiently.

 **TIP** At 3000 rpm, in a 4-cylinder engine, it does this about 100 times a second. Older systems are not as accurate as modern systems.

IGNITION SYSTEM MAINTENANCE

RMP This must be undertaken at manufacturers' recommended intervals. It is necessary to undertake maintenance on a vehicle's ignition system.

In small groups examine technical data for a number of different vehicles and list which components require maintenance and at what intervals. Present your findings to the rest of your group.

Vehicle make/ model	Type of ignition system	Components 1 Type of maintenance	Components 2 Type of maintenance	Components 3 Type of maintenance

List the ignition component(s) on an older vehicle's ignition system (distributor type) that will require maintenance:

- _____
- _____
- _____

List the ignition component(s) on a modern distributorless (DIS) vehicle's ignition system that will require the **most** maintenance:

- _____

PROCEDURE FOR SCHEDULED IGNITION MAINTENANCE

What needs to be undertaken before removing components from the ignition system?

_____

Before removing the high tension (HT) leads it is important to note their correct position. How is this achieved?

_____

_____

Once the leads have been removed what other checks can be undertaken?

_____

_____

Before replacing spark plugs it is important to always use the correct equipment. Name the specialist equipment required:

1 _____

2 _____

3 _____

When the spark plug has been removed which parts of it should be checked for damage?

_____

_____

What other components within the ignition system need to be checked?

_____

_____

Before refitting the spark plug which parts need to be set to manufacturers' specification?

_____

_____

What checks need to be carried out before starting the engine?

_____

_____

_____

 **RMP** Once you have completed this procedure you should start the engine to check the engine operation and the throttle response is correct.

TIP You need to be careful when carrying this out because it requires the engine to be running so there are hazards such as rotating parts and hot components. Also, remember that ignition systems operate at high voltages.

With the knowledge gained in this section undertake a full ignition service to include spark plug, HT leads, LT leads, distributor cap, arm and coil.

Note any readings or damage and compare it to the manufacturers' specification and make recommendations.

Vehicle make/ model	Spark plug type condition	HT lead type condition	LT lead type condition	Distributor cap type condition	Distributor arm type condition	Coil type condition

Multiple choice questions

Choose the correct answer from a), b) or c) and place a tick [✓] after your answer.

1 **An incorrectly adjusted electrode gap can cause an engine to:**

a) Misfire [ ]

b) Have no effect [ ]

c) Be more efficient [ ]

2 **What test can be undertaken to check the vehicle's emissions after an ignition system service?**

a) Brake roller test [ ]

b) Pressure test [ ]

c) Emissions test [ ]

3 **The purpose of an ignition coil is to:**

a) Increase voltage output [ ]

b) Decrease voltage output [ ]

c) Increase voltage to the battery [ ]

4 **A distributorless ignition system does not use:**

a) Coils [ ]

b) Mechanical contact breakers [ ]

c) High tension leads [ ]

5 **When checking a component for high resistance your multimeter should be set to:**

a) Ohms [ ]

b) Volts [ ]

c) Amps [ ]

SECTION 5

Introduction to spark ignition fuel systems

USE THIS SPACE FOR LEARNER NOTES

_____

_____

_____

_____

_____

Learning objectives

After studying this section you should be able to:

● Know how to work safely on spark ignition fuel systems.
● Be familiar with the components of a spark ignition fuel system.
● Understand the procedure for changing spark-ignition system air filters.
● Understand environmental considerations relating to spark ignition fuel systems.

Key terms

Tank   Used to safely store fuel for the engine.
Fuel filter   Prevents any dirt in the fuel reaching the carburettor/injectors.
Low pressure pump or lift pump   Used to pressurize the fuel injection system fuel lines/pipes used to transport safely the fuel to and from the tank.
Carburettor   A mechanical device for mixing fuel with air.
Stoichiometric   Chemically correct ratio of fuel and air required for complete combustion.
Single point   A single injector system which sprays fuel for all cylinders into the air at one place, usually by the throttle body in the inlet manifold.
Multi-point   An injection system in which each cylinder has its own injector. Only air enters the inlet manifold. The injectors are situated in the inlet manifold close to the valve ports.
ECU (ECM)   Electronic control unit (module).
Hydrocarbon (HC)   Causes respiratory problems, liver damage and cancers.
Oxides of nitrogen (NO_x)   – can cause respiratory conditions, smog and acid rain.
Carbon monoxide (CO)   Colourless, odourless, poisonous to human and animal life.
Carbon dioxide (CO_2)   Greenhouse gas that contributes to global warming.
Sulphur dioxide (SO_2)   Pollution and acid rain.
PM soot particles   Cause respiratory problems and cancers.

http://www.aa1car.com/index_alphabetical.htm

http://www.motor.org.uk/

Hazardous substances

Many liquids and substances used are either:

- Toxic
- Corrosive
- Irritants
- Harmful

Ensure you are aware of the correct precautions to take when working with hazardous substances. Control of Substances Hazardous to Health (COSHH) safety data sheets should be made available for all of these substances and they will state the precautions and actions to take.

CAUTION: Do not connect pressure testing equipment to the fuel rail or other high pressure fuel system components of an injection system. **Pressures can exceed 30000 psi.**

DO NOT attempt to slacken any high pressure fuel pipe connections with the ignition on or the engine running.

AVOID the risk of fire – NEVER work on the petrol injection system when SMOKING or close to a NAKED FLAME.

ALWAYS keep a fire extinguisher close at hand when working on the petrol injection system.

ALWAYS ensure that any replacement fuel system parts are correct for the application in question. Many units share common external features, but differ internally.

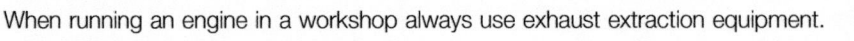

When running an engine in a workshop always use exhaust extraction equipment.

Remember: when working with fuel systems you may be dealing with:

- Very hot surfaces
- Volatile and flammable fuel under pressure
- Chemicals which are carcinogenic or toxic
- Rotating components
- Waste disposal

Always drain petrol from tanks by using a fuel retriever. For information on safety go to **http://www.hse.gov.uk/mvr/priorities/fire.htm**

Complete the following word search which includes key spark-ignition fuel system terminology.

```
L K I F F L T T N L F R P C
N F E C U B P L E T C M C G
C I R T E M O I H C I O T S
L E U N L A I F R L N N I L
E R S I F C L T R A E N I F
E T S O I B L P I E G C V A
N R E P L S M U K L O K O E
F P R I T F T M E I N N L E
R L P T E I T P E A I R A F
S L W L R L O C T S C A T N
R T O U I I F O I E R O I C
I N L M N F L A M M A B L E
R O T T E R U B R A C F E L
U N P E I O E U I O P A L M
```

MULTIPOINT   SINGLEPOINT   CARBURETTOR
LOWPRESSURE   LIFTPUMP   STOICHIOMETRIC
FUELFILTER   TANK   ECU
FLAMMABLE   VOLATILE   CARCINOGENIC

PETROL FUEL SYSTEMS

To enable a spark ignition engine to operate efficiently it requires the correct mixture of air and fuel. In the past, the mixing of petrol and air was carried out by using a carburettor. These are still found on some motorbikes and mowers. It uses the air entering the engine via the carburettor to carry a set amount of fuel into the combustion chamber.

Why are carburettors not used any more?

_____

Complete the following paragraph using words from the following word bank (note: there are two distracter words):

electronic	**injector**	**directly**	**spark**
manual	**fuel injection**	single point	

Modern engines use a _____ system that uses an injector(s) controlled by an _____

control unit (ECU). These injection systems use either one _____ per cylinder (multi-point

systems), or just one injector for all of the cylinders (_____). Some manufacturers even inject

the petrol _____ into the combustion chamber (direct injection).

The ECU monitors the operating conditions of the engine and attempts to provide an ideal air/fuel ratio 'stoichiometric'.

What is the ideal air/fuel ratio?

_____

Label the fuel system components in the diagram opposite. Add arrows and label the following list of components on the diagram:

fuel return pipe	**charcoal canister**	fuel tank	**fuel pump**
fuel vapour pipe	**fuel injectors**	fuel filter	
fuel feed pipe	**two-way valve**	**pressure regulator**	

Name the fuel system components shown below and on page 110:

_____

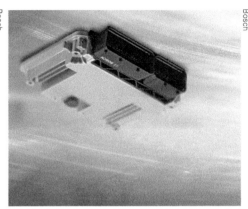

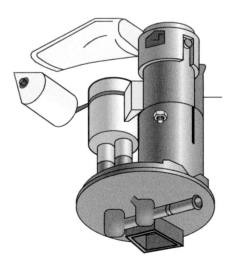

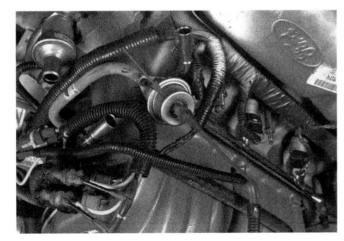

In small groups research the components listed on the previous page (one per group). You should include its location and what function it performs in a petrol fuel system. Present your findings to the rest of the group.

CHANGING A SI ENGINE AIR FILTER

The picture below shows a basic air induction system for a modern vehicle. Label the components indicated.

Honda

Place the titles and descriptions below in the correct order to indicate the correct procedure for replacing a vehicle's air filter.

☐ **Identify air filter:** It may be necessary to refer to the vehicle's service manual to identify the location and type of air filter fitted.

☐ **Clean debris off old air filter and housing:** Remove any dirt from around the air filter and housing as this could enter the system and damage the internal parts of the engine. Check the replacement filter is correct. Mark all intake pipes on the filter and housing before you disconnect the old ones, these may be electrical.

☐ **Installation:** Fit the new filter and secure all induction hoses and tighten the clamps. NOTE: ensure all connections are airtight as not to allow contamination into the system as this will cause premature wear. Check the filter is fitted correctly.

☐ **Ensure the correct equipment is used to remove old air filter:** Loosen all clamps on the induction pipes, filter and housing. It may be necessary to remove some other auxiliary equipment to gain access.

☐ **Immobilize:** Immobilize the vehicle and remove ignition keys.

How are any old components disposed of?

_____

_____

Carry out an air filter service on a range of workshop vehicles. Ensure you have all technical information before you start.

http://www.autodata.ltd.uk/

TIP

When you have removed the old air filter always compare it to the new one to ensure it is exactly the same. If it is not, contamination will get into the engine and this will cause premature wear.

FUEL SYSTEM MAINTENANCE

Before removing and replacing a fuel filter a technician must adhere to some basic workshop practices. Place the following statements in order of priority **before** the filter can be removed:

Remove all dirt from around the filter and housing

Depressurize the fuel system

Immobilize the vehicle

_____

_____

Once any work has been completed on a vehicle's fuel system what checks should a technician undertake?

_____

With regards to health and safety, petrol is said to be extremely:

_____

When changing a fuel filter you will have to disconnect the fuel pipe. What would you use to prevent fuel flowing from the pipe?

_____

What could be used to prevent any spilt fuel coming in contact with the workshop floor?

_____

TIP

It is important to service both the air and fuel systems to maintain clean air and fuel in the system. This will ensure that the engine runs efficiently and maintains the correct vehicle emissions and economy.

Multiple choice questions

Choose the correct answer from a), b) or c) and place a tick [✓] after your answer

1 **When a fuel filter becomes dirty what affect will it have on the engine?**

   a) The power will increase as the engine revs decrease [ ]

   b) It will restrict the quantity of fuel being delivered to the engine and will affect the vehicle's emissions and performance [ ]

   c) No effect is noticed on the engine under normal operating conditions [ ]

2 **When a paper element type air filter becomes dirty it should be:**

   a) Replaced [ ]

   b) Washed out in solvent [ ]

   c) Blown out with compressed air [ ]

3 **Name the component that prevents any dirty air reaching the engine inlet manifold:**

   a) Air filter [ ]

   b) Fuel filter [ ]

   c) Fuel pump [ ]

4 **What component forces the fuel to flow from the tank to the carburettor/injectors?**

   a) A flow restrictor valve [ ]

   b) Fuel pump [ ]

   c) Fuel rail [ ]

5 **Which component or components mixes the air and fuel in the correct percentages?**

   a) Inlet manifold [ ]

   b) Carburettor/injectors [ ]

   c) Exhaust gas recirculation valve [ ]

6 **One of the biggest differences between the petrol and diesel engine is how the fuel is ignited. Describe how this takes place in the SI engine.**

   a) Air only is drawn into the combustion chamber and ignited by a high voltage spark [ ]

   b) The air/fuel mixture is drawn into the combustion chamber and ignited by a high voltage spark [ ]

   c) Fuel is injected into a separate swirl pot and ignited spontaneously by the hot gasses [ ]

7 **Name the two most common methods that are used to inject petrol into a SI engine:**

   a) Inlet and exhaust valve [ ]

   b) Single point and multi-point injection [ ]

   c) Common rail and rotary pump [ ]

8 **What is the 'stoichiometric mixture ratio'?**

   a) 14.7 parts of air to 1 part of fuel [ ]

   b) 14.7 parts of fuel to 1 part of air [ ]

   c) 16.1 how hard the air is worked [ ]

SECTION 6

Introduction to compression ignition fuel systems

USE THIS SPACE FOR LEARNER NOTES

_____

_____

_____

_____

Learning objectives

After studying this section you should be able to:

- Understand how to work safely on compression ignition fuel systems.
- Identify the components of compression ignition fuel systems.
- Remove and refit compression ignition system filters.
- Understand environmental considerations relating to compression ignition fuel systems.

Key terms

Fuel filter Filters any contamination entering the fuel system.
Injection pump Increases the pressure of the diesel sufficiently for injection.
Injector Injects diesel into the combustion chamber under very high pressure.
Tank Stores the vehicle's fuel in a safe and controlled manner.
Glow plug Heats the air in the combustion chamber to aid cold starting.
Battery Used to power all of the electrical components located within the fuel system.
Crankshaft position sensor (CPS) Designed to signal the engine management system when the engine is in the correct position to run.
Engine control unit (ECU) The electronic brain of the vehicle where all sensors send inputs to monitor the engine efficiency.
Coolant sensor Senses the temperature of the vehicle's coolant.
Exhaust gas recirculation valve (EGR) Controls the amount of exhaust gasses that are recirculated back into the engine.

CAUTION: Do not connect pressure testing equipment to the fuel rail or other high pressure fuel system components of a common rail diesel injection system.

DO NOT attempt to slacken any high pressure fuel pipe connections with the ignition on or the engine running.

AVOID the risk of fire – NEVER work on the diesel injection system when SMOKING or close to a NAKED FLAME.

ALWAYS keep a fire extinguisher close at hand when working on the diesel injection system.

ALWAYS ensure that any replacement fuel system parts are correct for the application in question. Many units share common external features, but differ internally.

When running an engine in a workshop always use exhaust extraction equipment.

Remember: when working with fuel systems you may be dealing with:

- Very hot surfaces
- Volatile and flammable fuel under pressure
- Chemicals which are carcinogenic or toxic
- Rotating components
- Waste disposal

Always drain diesel from tanks by using a fuel retriever

For information on safety go to:

- **http://www.hse.gov.uk/mvr/priorities/fire.htm**
- **http://rb-kwin.bosch.com/gb/en/automotivetechnology/overview/ index.html**
- **http://auto.howstuffworks.com**

Complete the following word search which includes key compression ignition system terminology:

```
C U M . N T S C J J T A S L B R F O N Y E T O F A
T P S N R E T C L P C T T T T T S I S I I A I S T
T P L I I N A C O Y C U T T N R P E C O W O A P E
L N P R U N C C I N U R G E T F S O E R S E A N N
N J . S T S T K R F E L E U S N I L K I T P T E U
L E O F R U J E G S N O C E E E N I E P J S N E L
C U I S N S E O G S T N R A P R E S C U R W C R T
T J I A Y . A F B N C G H O P E T L A O B H E N E
N C S U T S T A O O E I E E P T O T A G C S O F A
U M T P E O T E N O R C S M G P E O I W E A S N Y
E R E . S T C R O U L S U I A L E F A I T O N O J
R G G J E R H E N G N P R O A P T E E E T H S T F
P F E R S I F L C F N U S I A U S S M O U R N E N
P S Y O U R N O I O R L C T E S A O S T S O A R P
U K O S L O E W I L F U U T M I R A E K N O O B S
Y I K N A T E T G C S T N S I C T U O O L E C P A
K O U E N C C L L . K T R R I G N J F A S I T R E
. R O S N E S N O I T I S O P T F A H S K N A R C
T I N T J J B S W S F N R N A T C N T L P G C E F
S T T N A N S T P J O L R T T I K T W N O E T P S
J Y I A N I P N L G R G E N E E I U I N N G T I G
A A L L Y O I O U T E T O U R N S E F S N E W T W
J P S O . P N I G A O U R C F I T F A O E A F H T
S T G O E N N L E T L B I L H N G S A L E S R T C
E P C C I U I N T S A A C C N S T U ⊥ L S N B O G
```

FUEL FILTER
ECU
CRANK SHAFT POSITION SENSOR
COOLANT SENSOR
BATTERY
GLOW PLUG
TANK
INJECTOR
INJECTOR PUMP
EGR

FOUR-STROKE COMPRESSION IGNITION (CI) ENGINE

A four-stroke diesel engine (CI) works exactly the same as a four-stroke spark ignition (SI) engine. It is in the combustion process where the main differences occur.

Complete the following paragraphs explaining the combustion process of a four-stroke compression ignition engine. Use the word bank below (note: there are two extra distracter words):

Inlet	fuel/air	injector	ignition
air	intake	atomize	
spark	compression	self-ignition	
directly	spontaneously	vapour	

During the _____ stroke, the diesel engine only sucks _____ into the combustion chamber through the _____ valve, **not** a mixture of _____. A diesel engine has no _____ plug; it is fitted with a high pressure _____. The job of the injector is to _____ the fuel.

The diesel engine works on _____: as the piston reaches the top of the _____ stroke, the air is highly compressed and very hot (around 1000°C). The fuel is injected _____ into the combustion chamber and the heat and pressure _____ ignites the diesel.

 **TIP** The compression ratio for a petrol engine is around 10:1 to 12:1. A diesel engine has a compression ratio of around 14:1 and can be as high as 24:1.

COMPONENTS OF A DIESEL SYSTEM

The main components of a diesel fuel injection system are shown below. Correctly label the components shown.

_____   _____   _____

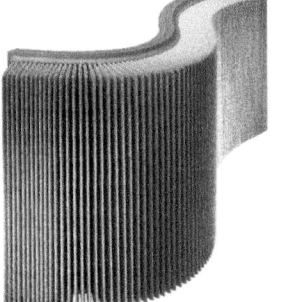

_____   _____   _____

All of the above images are supplied by Bosch

Bosch

The diesel fuel system comprises a low pressure system and a high pressure system.

Name the components indicated in the basic (CI) diesel injection systems from the list of terms.

Lift pump   **Injection pump**   **Injector pipe**
Primer   **Low pressure fuel pipe** **Return pipe**
Injector   **Leak off pipe**
Fuel filter   **Agglomerator**

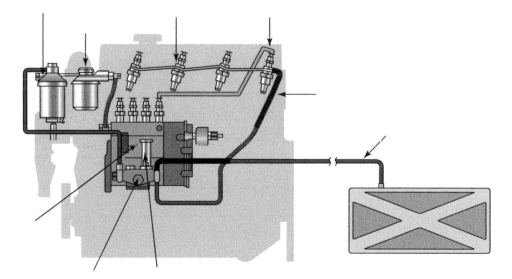

Name the type of fuel injector pump shown in the diagram opposite.

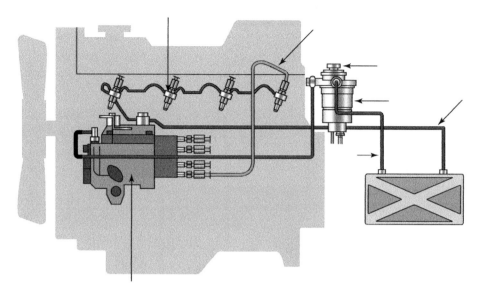

Name the type of fuel injector pump shown in the diagram above.

WWW **www.bosch.co.uk/content/language2/html/945.htm**

One of the biggest differences between the petrol and diesel engine is how the fuel is ignited. Describe the complete cycle of a four-stroke diesel (CI) engine.

_____

_____

_____

_____

Complete the following paragraph using the words from the wordbank below (note: there are two extra distracter words):

fuel	**injector**	**inlet valve**	**pressure**
cylinder	**tank**	**pipes**	**sedimenter**
sensor	**chamber**	**pump**	**leak-off**

In a basic diesel fuel system, the fuel _____ stores the diesel fuel. A lift _____ draws

the fuel from the tank. A _____ filter removes any water, and larger particles in the fuel. A

fuel _____ pump delivers fuel under very high _____ to the injectors, along injector

_____. An injector in each _____ sprays _____ into each combustion

_____. _____ pipes take unused fuel back to the tank.

MODERN DIESEL SYSTEMS

Diesel engine design has changed significantly over the last 10 years. This has been due mainly to the increased environmental concerns. The introduction of emissions legislation by the European parliament has forced manufacturers to use electronic fuel injection and specialist exhausts and electronics.

Common rail

Complete the following paragraph:

This is the most popular diesel fuel injection system used by many manufacturers. It uses the

basic principle of lifting the diesel from the tank using a _____, then uses a ___

_____ to feed electronically controlled injectors via a common fuel rail.

The below diagram shows a diesel engine complete with its common rail fuel injection system.

Indicate on the drawing the location of the components listed below:

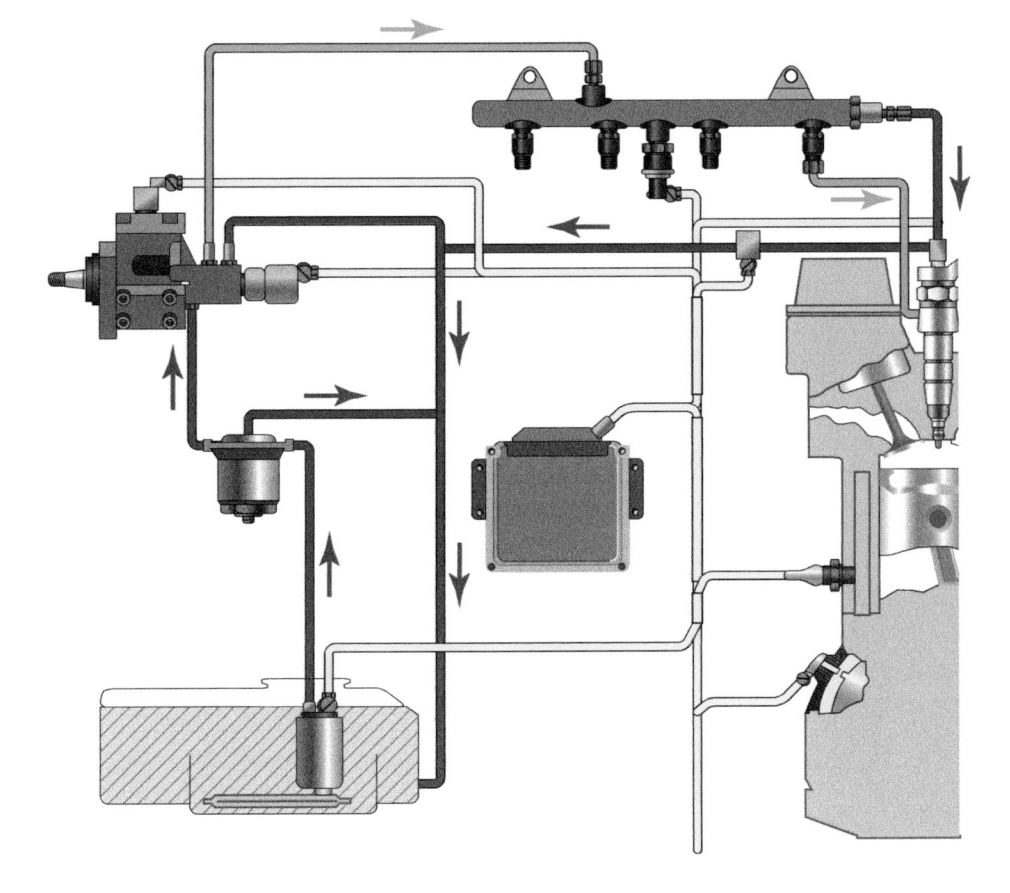

1	**Fuel tank**	8	**Fuel rail**
2	**Pick-up pipe**	9	**Fuel injector**
3	Low pressure pump	10	**Return pipe**
4	Fuel filter	11	**ECU**
5	**Low pressure pipe**	12	Crankshaft **position (CKP) sensor**
6	**High pressure** injection pump	13	**Coolant temperature sensor**
7	**High pressure pipe**		

Complete the following paragraph using the words from the wordbank below (note: there are two extra distracter words):

breather	**time**	**combustion**	**30 000**
pump	**injectors**	**pipes**	
delivery valve	**sensors**	**electronically**	
ECU	**rail**	**pressure**	

The high-pressure _____ increases the fuel _____ in the accumulator or fuel

_____ up to 2200 bar or _____ psi. The pressure is controlled by the _____

and the many different _____ on the vehicle. The fuel is then transferred through rigid

_____ to the _____ controlled fuel _____, which inject the correct amount of fuel

into the _____ chambers at the correct _____.

Some modern common rail injection systems operate at voltages of well over 100 volts and fuel pressures of up to 30000 psi.

TIP Do not 'feel' for leaks with your hands around the engine or loosen high pressure injection pipes.

Bosch

Discuss in small groups the correct procedure for the replacement of a new fuel filter and how any old components and fuel must be disposed of. Below are the main areas for discussion.

Descriptions for the procedure are below. List these in the correct order and explain how each action should be carried out.

Ensure the correct equipment is used to remove old fuel filter

_____

_____

_____

☐ PPE VPE

_____

_____

☐ Identify fuel filter

_____

_____

☐ Installation

_____

_____

☐ Final part

_____

_____

☐ Bleed system

_____

_____

_____

☐ Depressurize the fuel system

_____

_____

☐ Clean debris off old fuel filter

_____

_____

_____

☐ Immobilize

_____

With the knowledge gained in this section undertake an air and fuel filter service on a range of workshop vehicles. Ensure you have all the technical information required before you start.

WWW http://www.autodata.ltd.uk/

Multiple choice questions

Choose the correct answer from a), b) or c) and place a tick [✓] after your answer.

1 **Name the type of fuel injection system used on modern diesel cars:**

   a) Common rail [ ]

   b) Fixed rail [ ]

   c) Variable rail [ ]

2 **Name the components that operate on the high pressure side of a diesel common rail system:**

   a) Rail, injector pipes, injectors [ ]

   b) Tank, fuel filter, lift pump [ ]

   c) Return rail, fuel cooler, breather [ ]

3 **An incorrectly fitted air filter will cause an engine to:**

   a) Overheat [ ]

   b) Wear prematurely [ ]

   c) Overrun on shutdown [ ]

4 **On a rotary fuel injection pump the injector pipes are fitted:**

   a) On the back of the pump in a rotary shape[ ]

   b) On the top of the pump in a rotary shape [ ]

   c) On the front of the pump in a rotary shape [ ]

5 **What is the normal operating pressure of a common rail high pressure pump?**

   a) 400 to 800 bar [ ]

   b) 800 to 2200 bar[ ]

   c) 1350 to 3500 bar [ ]

6 **Unused diesel fuel is returned to:**

   a) Injector pump and fuel tank [ ]

   b) Low pressure pump and fuel tank [ ]

   c) Injectors and fuel tank [ ]

PART 3 CHASSIS

USE THIS SPACE FOR LEARNER NOTES

SECTION 1
Vehicle construction 124

SECTION 2
Suspension and steering systems components and maintenance 131

SECTION 3
Vehicle braking systems components and maintenance 147

SECTION 4
Vehicle wheels and tyres construction and
maintenance 159

SECTION 1

Vehicle construction

USE THIS SPACE FOR LEARNER NOTES

_____

_____

_____

_____

_____

_____

_____

Learning objectives

After studying this section you should be able to:

- Identify different vehicle drive layouts.
- Identify body parts on light vehicles.
- Identify trim components found on light vehicles.
- Identify different body types (also covered in Part 1 Section 5).

Key terms

Clutch Provides a temporary position of neutral and enables a smooth take-up of drive.

Gearbox A major unit which multiplies the engine torque (turning force) and provides a means of reversing the vehicle, as well as a permanent neutral.

Final drive Takes drive from the gearbox to the driven wheels.

In-line An engine layout where the cylinders run from front to back.

Transverse An engine layout where the cylinders are positioned from side to side (across the vehicle).

FWD Front-wheel drive.

RWD Rear-wheel drive.

AWD All-wheel drive.

4WD Four-wheel drive.

DRIVE LAYOUTS

Vehicles have various drive layouts and wheels are driven in a variety of combinations.

Drive layouts fall into one of three main categories:

- **Front-wheel drive (FWD)**
- **Rear-wheel drive (RWD)**
- **Four-wheel drive (4x4)**

Using the diagrams which follow identify each layout shown from the list below. Give an advantage of each layout and an example of a vehicle that uses each layout:

- **In-line front engine, rear-wheel drive with a Solid rear axle**
- **Transverse front engine, front-wheel drive**
- **In-line rear engine, rear-wheel drive**
- **Transverse mid-engine, rear-wheel drive**
- **Front in-line engine, four-wheel drive**
- **In-line front engine, rear-wheel drive with independent rear suspension**
- **In-line mid-engine, rear-wheel drive**
- **Transverse front engine, four-wheel drive**

Label the drive line components in the speech bubbles using the following terms:

Clutch	Gearbox	Transfer box	Live axle
Final drive	**Propeller shaft**	**Gearbox including**	
Engine	**Driveshaft**	**final drive**	

Type of layout: _____

Advantage of layout: _____

Example model: _____

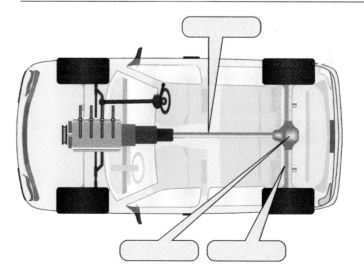

Type of layout: _____

Advantage of layout: _____

Example model: _____

Type of layout: _____

Advantage of layout: _____

Example model: _____

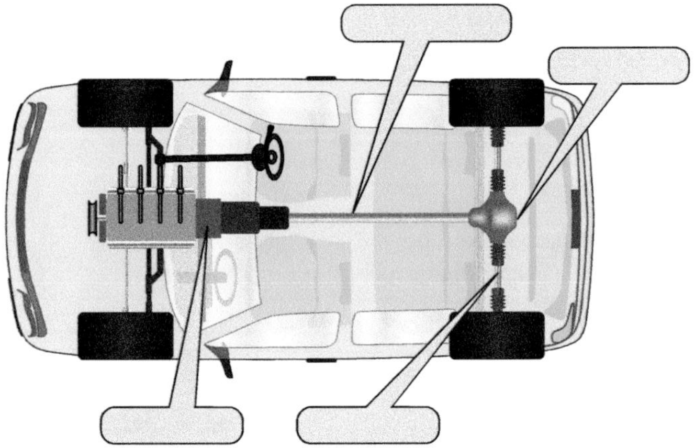

Type of layout: _____

Advantage of layout: _____

Example model: _____

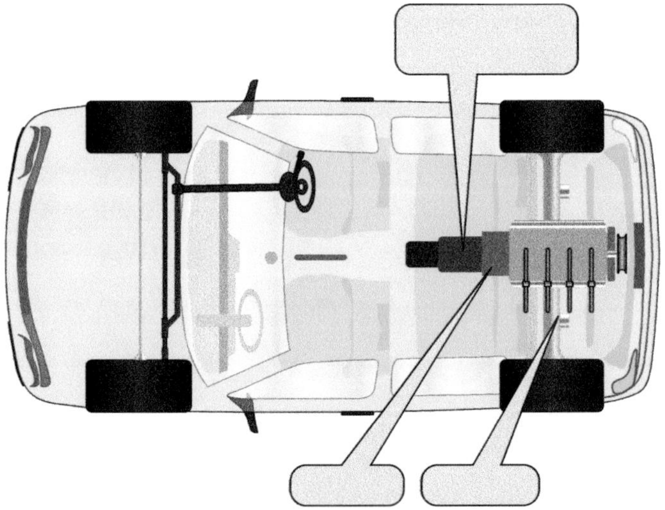

Type of layout: _____

Advantage of layout: _____

Example model: _____

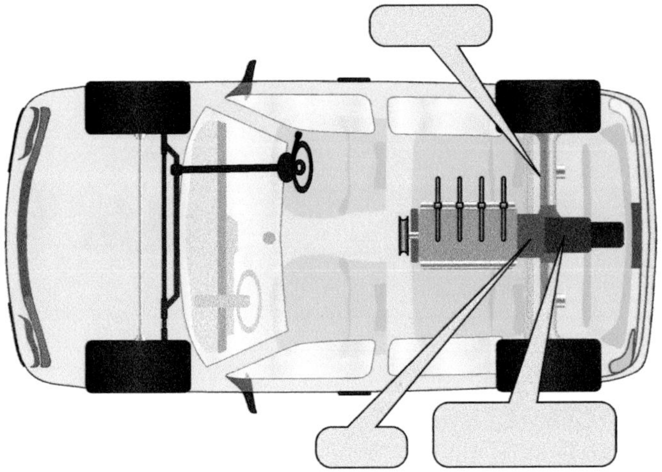

Type of layout: _____

Advantage of layout: _____

Example model: _____

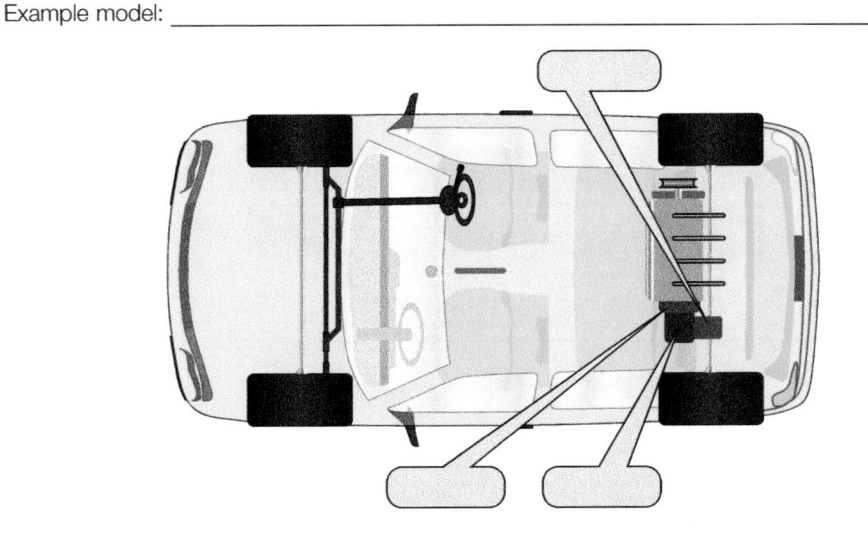

Type of layout: _____

Advantage of layout: _____

Example model: _____

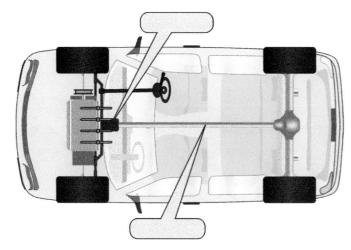

Type of layout: _____

Advantage of layout: _____

Example model: _____

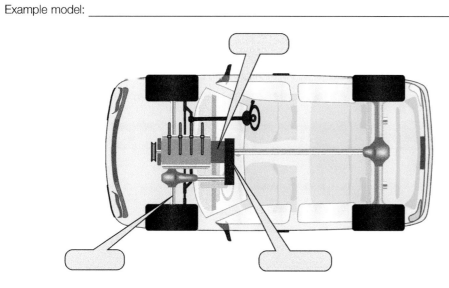

In your workshop ask your supervisor to safely raise the following types of vehicle on a vehicle hoist.

- Front engine with front-wheel drive
- Front engine with rear-wheel drive

Draw the above layouts using the relevant components from the list:

Engine	**Final drive**	**solid axle**
Clutch	**Driveshafts**	**Exhaust system**
Gearbox	**Propeller shaft**	**Wheels**

Front engine with front-wheel drive:

Vehicle make – _____   Model – _____

Front engine with rear-wheel drive:

Vehicle make – _____   Model – _____

 Only use components applicable to your chosen vehicle.

NON-STRUCTURAL BODY PARTS

Label the parts listed below on the following diagram and photos:

| Bonnet | Boot lid | Door | Rear quarter panel |
| Wing | Tailgate | Front windscreen | |

MAIN TRIM COMPONENTS FOUND ON LIGHT VEHICLES

Determine the trim or body part being described using the list of components on page 129 to help you. Identify the part on a vehicle in your learning environment.

1 This part is operated electrically or manually. One function is to allow air into the vehicle for driver comfort.

_____

2 Trim used to finish the inside of the roof.

_____

3 Main type of driver or passenger restraints.

_____

4 Located at the front and rear of the vehicle, absorbing front or rear impacts.

_____

5 This item separates the boot space from the vehicle interior.

_____

6 Part of the vehicle interior where the instrument cluster is located.

_____

Choose components from the list below to name the items indicated in the following photos:

Headlining   Dashboard   Door drop glass   Rear lamp units
Airbag   Seat belts   Glove box   Front offside light
Parcel shelf   Bumper   Centre console   cluster
Door moulding   Tail light   Tailgate

Across

2. Name of the rear opening door of a hatchback.
5. Component fitted between the engine and the gearbox.
6. Fitted at both ends of the car to absorb an impact in the event of a front or rear end smash.
7. Engine configeration where the engine is positioned behind the rear wheels.

Down

1. Trim used inside the car next to the rear window glass (two words 6-5) - separates the boot from the interior.
3. Engine configeration that fits into the car from front to rear.
4. Inside trim where the speedometer is situated.

Multiple choice questions

Choose the correct answer from a), b) or c) and place a tick [✓] after your answer.

1 What is the meaning of this abbreviation: AWD?

a) All-wheel drive [ ]

b) Alternating wheel drive [ ]

c) Adjustable wheel drive [ ]

2 Name the component that connects and disconnects drive from the engine to the gearbox:

a) Final drive [ ]

b) Transfer box [ ]

c) Clutch [ ]

3 Which layout is most likely to give equal front/rear weight distribution?

a) Rear engine, rear-wheel drive [ ]

b) Front engine, front-wheel drive [ ]

c) Mid-engine [ ]

4 Identify the non-structural body part:

a) Chassis [ ]

b) Bonnet [ ]

c) Roof panel [ ]

5 The term used when the engine is fitted side to side in the vehicle:

a) In-line [ ]

b) Longitudinal [ ]

c) Transverse [ ]

SECTION 2

Suspension and steering systems components and maintenance

USE THIS SPACE FOR LEARNER NOTES

_____

_____

_____

_____

Learning objectives

After studying this section you should be able to:

- Work safely on steering and suspension systems.
- Identify non-assisted steering and suspension components.
- State how steering and suspensions systems operate.
- Undertake routine maintenance on steering systems.
- Undertake routine maintenance on suspension systems.

Key terms

IFS   Independent front suspension.
IRS   Independent rear suspension.
Live axle   Driven axle.
Dead axle   Non-driven axle (usually on the rear).
Beam axle   Rigid axle.
Spring   A suspension device designed to absorb road shocks, i.e. a spring.
Damper   A device which dampens the oscillations or vibrations of the road springs.
Bump   Upward movement of suspension.
Rebound   Downward movement of suspension.
Pitch   Forward and backward rocking motion of the vehicle.
Roll   Sideways sway or 'leaning outwards' of a vehicle on corners.
Panhard rod   A rod mounted between the body or chassis and the axle, to control the lateral (sideways) movement of the axle.
Sprung weight   All of the vehicle's mass that is above the spring.
Un-sprung weight   All of the vehicle's mass that is below the spring.

When you are raising and supporting a vehicle, be sure to do this safely and correctly.

Position the jack and axle stands at the correct place under the chassis. Chock the road wheels. When using a lift ensure that you position the vehicle correctly. Beware of overhead obstructions when raising the lift.

Components within a suspension system are loaded and under tension or compression when mounted or removed from the vehicle. Ensure you use an approved tool. Make sure that it remains properly secured to any component that you may be working on, especially if a spring is removed.

You must also:

- Be aware that alternative types of suspensions can contain liquids under high pressures or even air
- Always keep your work area tidy and uncluttered
- Adhere to all correct waste disposal procedures

www.pro-align.co.uk

www.motorera.com/dictionary

http://auto.howstuffworks.com

SUSPENSION SYSTEMS

The purpose of the suspension system on a vehicle is to minimize the effect of road surface irregularities on passengers, the vehicle and its load. The suspension system must also keep the tyres in contact with the road surface.

E	B	D	S	S	N	I	D	I	G	E	D
E	G	O	P	E	L	I	M	B	U	B	A
G	N	I	R	P	S	P	A	R	I	D	A
E	U	R	U	O	L	V	E	P	N	T	P
O	R	T	N	D	A	L	B	U	M	P	P
G	P	G	G	R	X	L	O	T	D	B	R
G	S	N	W	A	P	B	P	R	R	E	B
I	N	D	E	P	E	N	D	E	N	T	D
B	U	V	I	R	N	E	P	E	I	E	R
E	I	R	G	P	P	M	E	U	G	A	E
L	D	E	H	I	A	T	R	H	T	G	E
W	N	L	T	D	D	E	U	U	E	N	D

REBOUND
BUMP
DAMPER
BEAM
LIVEAXLE
ROLL
SPRUNGWEIGHT
UNSPRUNG
INDEPENDENT
SPRING

From the list below identify the main components that form the basis for any suspension system:

- tyres
- springs
- dampers
- steering wheel
- wheel rim

Many different forms of suspension are used on all road vehicles. Some are very simple and relatively inexpensive and others are highly sophisticated and expensive. Suspension systems fall into one of two main categories: **independent** and **non-independent** systems.

Identify from the list below the main types of materials that are used in the manufacture of suspension springs:

- steel
- rubber
- gas
- wood
- plastic
- iron
- aluminium

TIP

As the vehicle rides over a **bump**, the spring will be compressed. This is called 'bounce'. When the vehicle wheel drops into a hole the spring will extend, and this is called '**rebound**'.

Identify and name the five types of suspension springs shown in the images:

- Leaf spring (mono)
- Coil (helical)
- Torsion bar
- Airbag
- Leaf spring (multi)

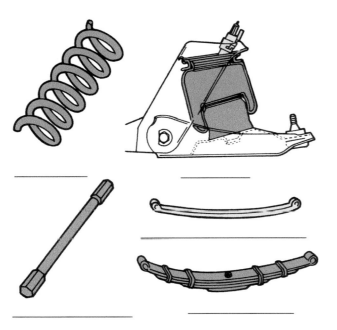

_____   _____

_____   _____

_____

Leaf spring

Leaf springs can still be found on some older car rear suspensions and are still used extensively in light and heavy commercial vehicles. They are used in conjunction with a **beam axle**.

Complete the paragraph below using the terms from the following word bank (note: there are two extra distracter words):

eye	hanger bracket	shackle
helper	main	non-independent
both	connected	spring

The leaf spring is a simple _____ suspension system. At _____ ends of the _____ leaf is a rolled _____. A bush is fitted into both _____ eyes and then the spring is _____ to the vehicle body or chassis via _____ pins.

Name TWO advantages and ONE disadvantage of a leaf spring.

Advantages:

1 _____

2 _____

Disadvantage:

1 _____

Label the multi-leaf spring suspension assembly shown using words from the list below:

Swinging shackle	Fixed shackle	Hanger bracket	Chassis
'U' bolts	Saddle	Main leaf	Spring pin
Centre bolt	Spring clip	Shackle pins	Axle

Why is the rear spring eye allowed to pivot on the chassis?

_____

_____

On the picture below, draw a leaf spring when a load has been exerted on it.

No load   Shackle

Complete the paragraph below using the terms from the following word bank. (note: There are two extra distracter words):

U	**axle**	**shackle**	**centre**
eyes	**increases**	**extends**	**spring**

The leaf spring is located on the axle by a _____ bolt and secured to the axle by _____

bolts. As the spring _____, the distance between the spring _____, that is, the length of

the spring, _____. This is accommodated by the swinging _____.

Torsion bars

A torsion bar spring is a straight metal bar that is fixed to the vehicle's body at one end and to the suspension control arm at the other end.

How does the torsion bar absorb wheel movements?

_____

_____

Label the diagram of a torsion bar setup shown using the words below:

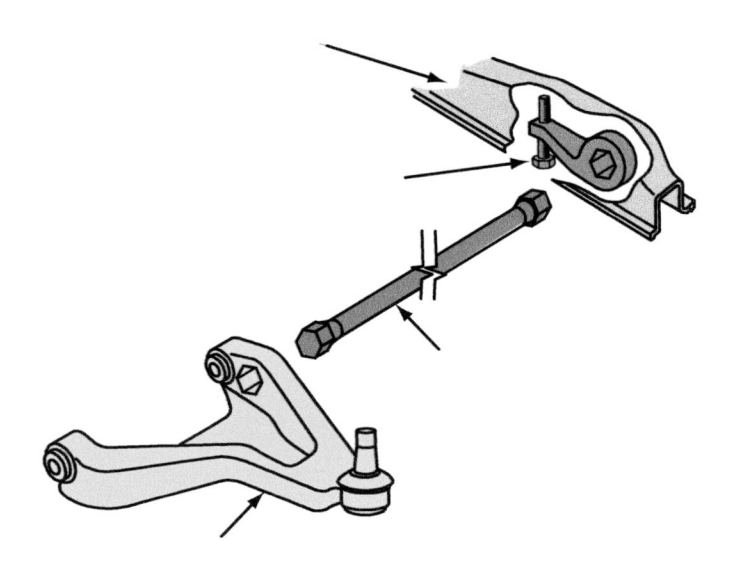

crossmember	torsion bar adjuster nut
torsion bar	lower control arm

TIP The torsion bar supports the vehicle's weight and load as it twists around its own centre and is placed under a torsional load.

How is the ride height or suspension stiffness of a vehicle fitted with torsion bar suspension adjusted?

_____

_____

Coil springs

Coil or helical springs are made from sprung steel rods heated and formed or shaped into a coil. The strength of the suspension spring can depend on the diameter and length of steel rod used or even its shape. This is decided by the manufacturer.

 Coil springs can look identical but may be very different. They can have different load capacities. There are often different colours painted onto the coils. These are codes for part identification.

Complete the paragraph below using the terms from the following word bank (note: there are two extra distracter words):

leaf	**soft**	**hard**	**suspension**
large	**light**	**heavy**	

The coil spring is used on many car front and rear _____ systems. When compared with the

_____ spring the coil spring is _____ in weight, provides a ride which is _____

and wheel deflection that can be _____.

 Caution! Coil springs exert a tremendous force. If these are to be removed make sure that the springs are contained with a spring compressor before dismantling systems.

Rubber springs

On modern vehicles the use of rubber in suspension systems tends to be in the form of 'bump stops' which prevent direct metal-to-metal contact.

Under what circumstances might the vehicle's suspension contact the chassis?

_____

_____

 Some commercial vehicles use rubber as the main springing material.

Rubber can also be used in conjunction with metal to form metalastic bushes. These are a combination of rubber and metal and are used in many suspension systems as well as engine mountings where movement and vibration need to be controlled.

Metallic bushes act as suspension fulcrum points, they allow for movement of the component they are securing, while maintaining its correct position.

Complete the paragraph below using the terms from the following word bank. (note: There are two extra distracter words):

rubber	**sleeve**	**bonded**
steel	**outer**	**movement**
plastic	**solid**	**independently**

The bush has a _____ outer casing and inner _____. The rubber is bonded to

both inner and _____ metal sleeves. As _____ occurs the inner sleeve can move

_____ of the outer sleeve via the _____ middle section of _____.

How are these types of bushes removed or fitted into components?

_____

_____

When used in suspension systems the metalastic bushes have to have a controlled amount of movement.

This movement is known as: _____.

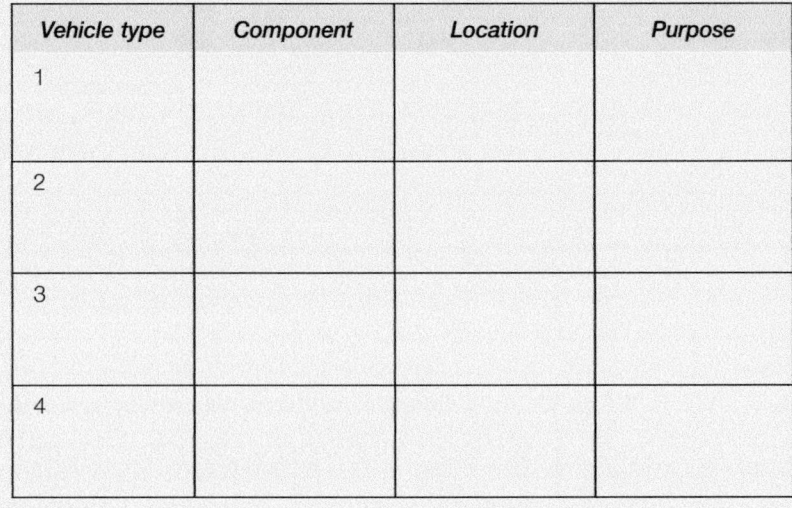

RMP   When inspecting metalastic bushes a technician should check for cracking and debonding. This is when the rubber and the metal separate.

Examine four different vehicles and list the location and purpose of any rubber components that are used in the suspension system.

Vehicle type	Component	Location	Purpose
1			
2			
3			
4			

Air suspension

An air suspension system usually consists of a reinforced rubber bag mounted between the axle and chassis. Air suspension provides a soft, cushioned ride and the ability to alter a vehicle's ride height.

What type of vehicle usually uses this type of suspension?

_____

The pressure in the bag can be adjusted electronically or mechanically using leveling valves.

Name one advantage of air suspension:

_____

_____

On the drawing below show where the airbag would be located.

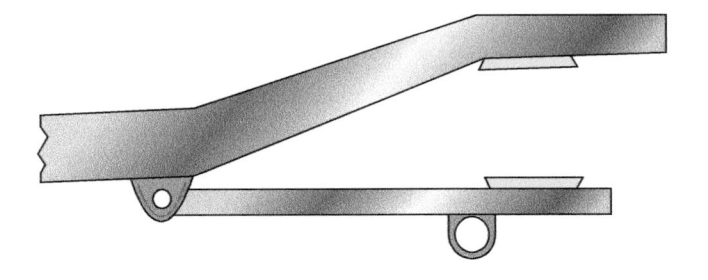

Suspension dampers

When the vehicle goes over a bump in the road the **spring** will flex as it absorbs the shock loading. If this is not controlled the spring will continue to flex for some time. If this is allowed to happen it will cause the vehicle to bounce which will result in a very uncomfortable ride and make vehicle road-holding difficult.

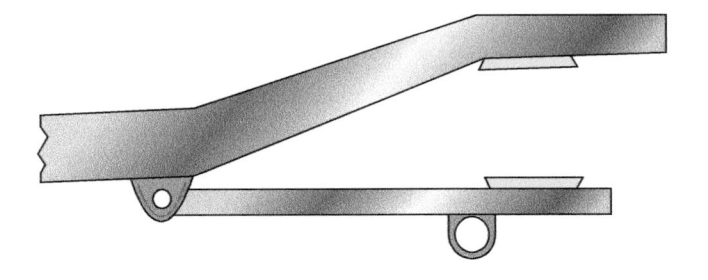

What is the function of the damper?

_____

_____

_____

The basic principle of hydraulic dampers is they pass oil through a series of small holes in a piston.

Shown opposite is a sectioned damper.

The flexing of the spring can also be called an oscillation. It will continue until all of the energy stored in the spring (known as kinetic energy) has been used.

What can be used inside a damper to control the amount of bounce of a spring?

_____ .

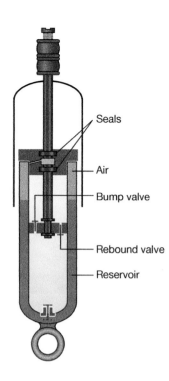

Seals

Air

Bump valve

Rebound valve

Reservoir

In small groups, examine a number of dampers. After you have inspected their condition, complete the table below.

Damper	Bushes	Cover	Piston rod	Seal	Damping	Serviceability
1						
2						
3						
4						

TIP The damper is sometimes incorrectly called a shock absorber.

Select a damper and hold it at each end. Try to operate it quickly which should be easy to do. Why is this?

_____

Dampers are normally replaced in pairs. When you check dampers, look at other components too – the chassis area, the springs, the suspension linkage, and so on. It is good practice to report any faults found. Don't just look at one thing – always watch out for other problems.

Complete the paragraph below using the terms from the following word bank. (note: there are two extra distracter words):

body	added	wheel	surface
ride	axles	weight	springs
heavy	adjustable	transferring	
height	maintain		

Suspension _____ are fitted between the vehicle _____ and the vehicle _____. This

allows the _____ to follow the uneven _____ of the road without _____ the

movement to the vehicle body. Springs are also designed to _____ a vehicle's _____

_____ and support its _____ even when a load is _____.

In pairs, locate a vehicle's coil springs. If they form part of a suspension unit draw the component and show how they are fitted into it.

On some vehicles the suspension coil spring can be located on the suspension strut (often known as a MacPherson strut).

What else is built into a Macpherson strut?

_____

What type of suspension system commonly uses a Macpherson strut?

_____

Remove and replace a Macpherson strut

The pictures below show the basic operation of removing a suspension strut. Place the pictures in the correct sequence of removal and assign the correct instructions to each picture.

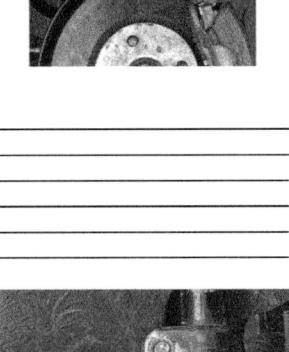

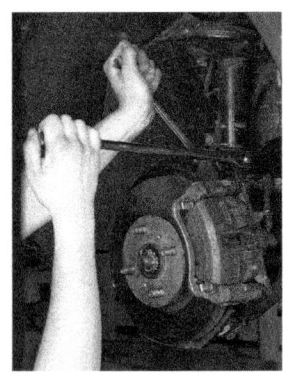

1 The top of the strut assembly is mounted directly to the chassis of the car.

2 Prior to loosening the strut chassis bolts, scribe alignment marks on the strut bolts and the chassis.

3 With the top strut bolts or nuts removed, raise the car to a working height. It is important that the car be supported on its frame and not on its suspension components.

4 Remove the wheel assembly. The strut is accessible from the wheel well after the wheel is removed.

5 Remove the bolt that fastens the brake line or hose to the strut assembly.

6 Remove the strut's two steering knuckle bolts.

7 Support the steering knuckle with wire and remove the strut assembly from the car.

8 Reinstall the strut assembly into the car. Make sure all bolts are properly tightened and in the correct locations.

WWW If you are unsure about any part of a task, information about removal and refitment or any technical data can be found at **http://www.autodata.ltd.uk/**.

www.youtube.co.uk can also be a useful guide.

Multiple choice questions

Choose the correct answer from a), b), c) or d) and place a tick [✓] after your answer.

1 **What would cause a vehicle to** pitch?

   a) Acceleration and braking [ ]

   b) Cornering hard [ ]

   c) Smooth gear changes [ ]

2 **What can be added to a suspension system to reduce body** roll?

   a) Anti-roll bar [ ]

   b) Damper [ ]

   c) Oversize tyres [ ]

3 **Where on a vehicle is un-sprung mass located?**

   a) Above the spring [ ]

   b) Below the spring [ ]

   c) Within the spring [ ]

4 **Which component on a vehicle's suspension system absorbs any road shocks first?**

   a) Coil spring [ ]

   b) Leaf spring [ ]

   c) Tyre [ ]

5 **A leaf spring fitted to a suspension system provides?**

   a) Axle location [ ]

   b) Support for the weight of the vehicle [ ]

   c) A means of controlling shock loading [ ]

   d) All of the above [ ]

STEERING

Key terms

Rack   Toothed bar.

Pinion   Gear wheel which engages with a rack.

Tie rod   Connecting rod or bar, usually under tension.

Track rod   Bar connecting the steering arms.

Drop arm   Connects steering box to drag link of steering system.

Idler arm   Similar to the drop arm but having a guiding function only.

Steering box   Changes rotary movement into linear movement.

Drag link   Connects drop arm to first steering arm.

Ackerman linkage   Form of steering arranged to give true rolling motion round corners.

Toe in or toe out   Inward or outward inclination of the leading edge of the front wheels.

Remember – when working with steering, you may be dealing with:

● Jacking and supporting vehicles

● High fluid pressures

● Loaded components under tension or compression

● Rotating components

● Airbags

● Waste disposal

http://www.aa1car.com/index_alphabetical.htm

www.carbibles.com/steering_bible.html

http://auto.howstuffworks.com/steering.htm

```
M K D C C P K I N K S T
  M R A G N I R E E T S
I R A C K M E K D E E T
R E G K E A C O A P E M
I D L E R A R M I A R A
O O I R R K D N T A I B
A R N M C N I O P R N E
N E K A T O E O U T G O
E I R N N I R N O E B E
E T T T N D O O D C O T
K D A X A R G G N   X O
R N R I N E   C K E R O
```

RACK
TOEOUT
ACKERMAN
TOEIN
STEERINGARM
STEERINGBOX
DROPARM
DRAGLINK
IDLERARM
TRACKROD
TIEROD
PINION
RACK

Steering system principles

The steering system has to provide a means of changing or maintaining the direction of a vehicle in a controlled manner. To steer the road wheels, the driver turns the steering wheel, this then uses a series of rods and gears which move the wheels.

The most popular steering system used on cars and light commercials is called the 'rack and pinion'.

The rack (a solid bar) slides inside a tube (the housing) and the pinion is the small gear mounted in a casing that meshes with the rack. The pinion is turned by the steering column that is connected to the steering wheel.

RMP

The most common faults in steering systems are:
- worn linkage ball joints
- stub axle swivel wear
- worn suspension ball joints
- worn steering gear
- oil leakage.

Below is a sectioned diagram of a rack and pinion. Name the numbered components on the assembly shown below:

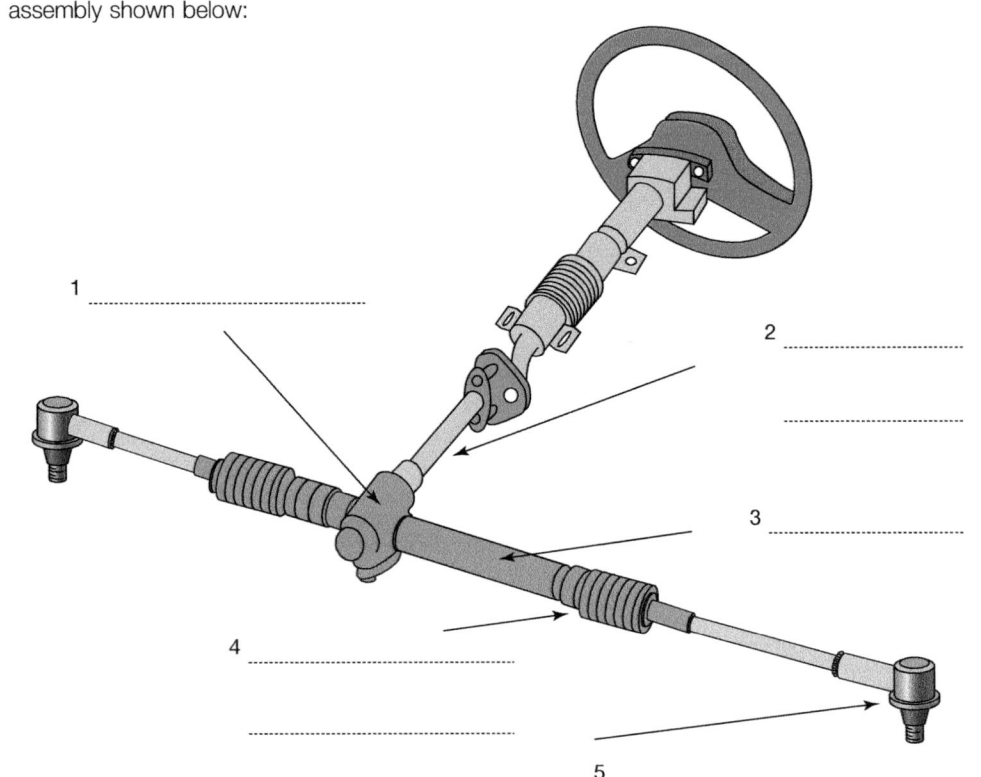

1 ....

2 ....

....

3 ....

4 ....

....

5 ....

....

What is used to reduce the wear in the steering rack?

_____

Apart from allowing for track rod movement, why are the bellows or gaiters fitted to the rack?

_____

How will making the pinion gear smaller affect the steering?

_____

_____

Steering box systems

The most popular type of steering box is the re-circulating ball. It can be manual or power assisted. The diagram below shows the internal components of this type of box. It is based upon the same principle as a 'nut and bolt'. A line of individual ball bearings are used to create the 'thread' inside the nut which runs around the worm shaft (bolt). These steering boxes are typically used on heavy goods vehicles, public service vehicles and some 4 × 4s.

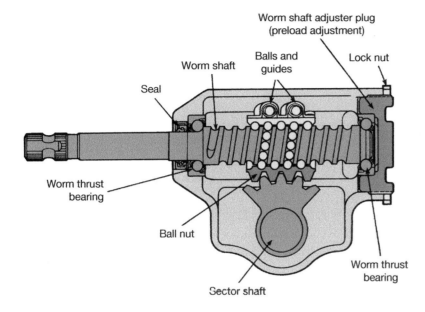

In small groups explain what happens when the steering wheel is turned and how the steering box operates. Drawings may assist in this process.

_____

_____

The diagram opposite shows a steering system that uses a steering gearbox. In small groups study the diagram. Draw in the steering linkage that joins the track rod arm to the steering arm and the drop arm.

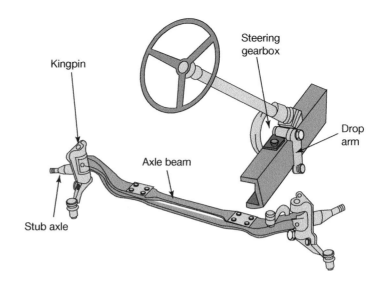

Explain how this system moves the steered wheels.

_____

_____

_____

_____

Ball joints or track rod ends

Ball joints are swivel connections threaded at one end onto the outer ends of the track control rods or track rod, while also being secured into the stub axle assembly or steering arms. The ball joints can swivel and rotate which allows the wheel to be steered even when it is moving up and down or even turning. They are threaded on to the rods as this allows for replacement when the ball joint becomes worn and also allows for the adjustment of the front wheel alignment (tracking).

The two main types of ball joints are shown below. Indicate on the diagram the names of the different areas of the joint.

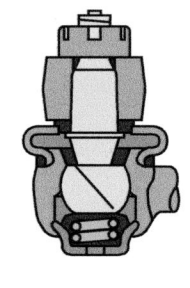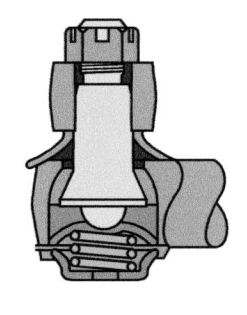

Name the tool that can be used when removing ball joints.

_____

http://www.sealey.co.uk/

RMP To check the linkage ball joints, get someone to 'rock' the steering wheel with the weight on the road wheels. Look for side-to-side movement or lift.

WWW If further guidance is required visit **www.vosa.gov.uk**

Apart from excessive movement what else would you check on the ball joints?

_____

TIP When you replace a track rod end, screw it on until you reach the same position as the old one. The lock nut can be used as a guide; it may also be necessary to mark it. If you do this the wheel alignment will need little or no adjustment, but always check the wheel alignment.

Steering column

In older cars, the steering column was a straight shaft, running inside a hollow tube. The steering wheel was attached to one end, and the steering box or rack to the other. In a collision, two forces were applied to the steering column; caused by the driver's mass hitting the steering wheel and the force of impact upwards from the collision. This would cause the driver very serious injury or even death.

To overcome this problem and to increase vehicle safety what do modern vehicles have built into their steering column?

_____

_____

What safety features are now built into all steering wheels?

_____

Power assisted steering (PAS)

A power steering system is used to reduce the driver's effort required when turning the steering wheel. It is of most benefit when parking or undertaking slow manoeuvres.

Power assistance can be provided by a high pressure hydraulic system or even an electrical motor. It must provide road feel and be accurate. It is important to understand that it is only designed to provide 'assistance' and should this fail there will still be a mechanical connection between the steering wheel and the road wheels, although the steering will feel heavier for the driver.

The diagram below shows a simple hydraulic power assisted steering system. Label the main components of the system using the list below.

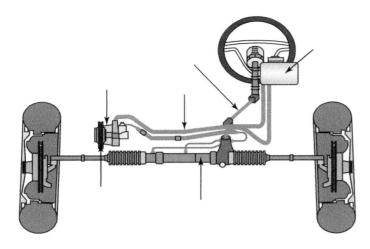

Hydraulic fluid and reservoir
Hydraulic pump

Drive belt pulley
Hoses

Power assisted rack assembly
Assembly steering column

State TWO advantages of power-assisted steering on the modern vehicle:

1 _____

2 _____

Discuss the type of maintenance that would be required for a hydraulic power assisted steering system.

_____

_____

_____

Electrical power assisted steering (PAS)

In an electrical power assisted steering system an electric motor is used to assist the driver, rather than a hydraulic system described above. The electrical motor is mounted directly onto the steering column and as the steering wheel is turned the motor is activated and assists the driver.

A basic electrical power assisted steering system arrangement is shown below. Label the main components on the diagram: **Worm and wheel radiation gear, Electric motor, Rack and pinion**

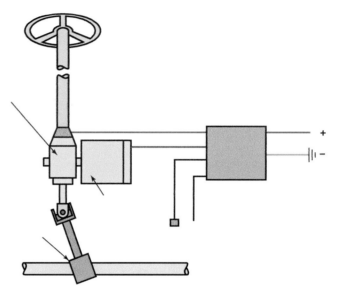

In small groups, discuss the advantages of this system over a pressurized fluid system.

_____

_____

_____

_____

What voltage is used to power the electric motor?

_____

FRONT WHEEL ALIGNMENT

The wheels on either side of a vehicle must be in line with each other when *rolling straight ahead*. That is, they must be parallel with each other.

The wheel alignment, or tracking, on the steered wheels is adjusted and set to the manufacturer's specifications. This setting varies from vehicle to vehicle. It depends on the vehicle's design and on the type of steering, suspension, tyres, transmission, and so on. They are set either toe in, toe out or parallel:

- *Parallel* **is when both wheels are perfectly aligned with each other.**
- *Toe in* **is when the front of the wheels are closer together than the rear of the wheels.**
- *Toe out* **is when the rear of the wheels are closer together than the front.**

On the two diagrams below indicate which set of wheels are toeing in and which are toeing out.

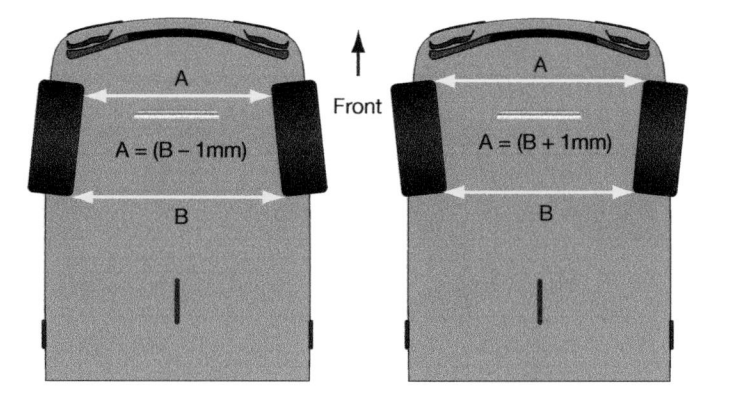

A = (B − 1mm)

A = (B + 1mm)

Front

_____   _____

State the main effects of the wheel alignment being set incorrectly:

1 _____

2 _____

3 _____

CHECKING AND ADJUSTING WHEEL ALIGNMENT

Before checking the wheel alignment or adjusting it, you should check the tyre pressures and condition of the rims. Also check the condition of all ball joints (linkage and suspension). A worn ball joint or a bent track rod will directly affect the wheel alignment.

 Always check manufacturer's data for exact settings of alignment.

The specialist workshop equipment used to check wheel alignment is shown below and on page 145. Identify and name the THREE different types of tracking gauges shown:

Sealey

_____

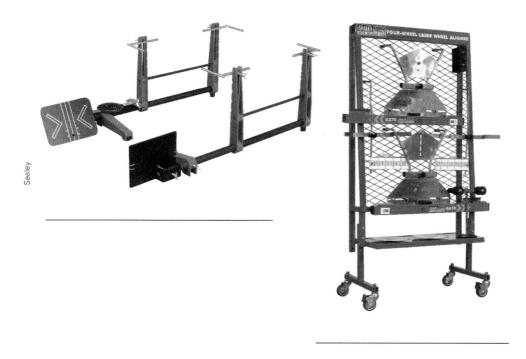

Sealey

It is always necessary to calibrate the gauges before you start. In small groups discuss why this may have to be undertaken.

Ackerman's principle is the layout or design of the steering linkage. It requires the distance between the track rod ends to be shorter than the distance across the steering axis swivels (wheel to wheel). This helps to automatically centralize the steering wheel when it is released after being held on a lock. The angles of the steering arms dissect at the centre of the rear axle.

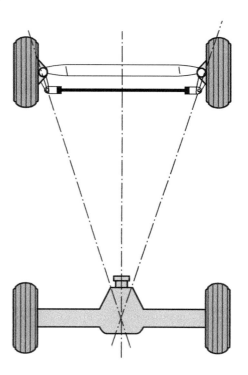

It also allows the inner wheel to turn through a larger angle when the steering is turned.

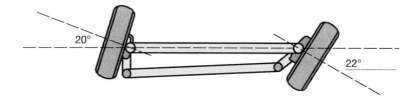

20°

22°

145

You should be familiar with the other angles that are used in a vehicle's steering system. In small groups investigate:

Swivel pin inclination   **castor and camber**

	Angle	*Where found*	*Typical quantity*
Castor and camber			
Swivel pin inclination			

Choose a workshop vehicle and check the tracking (alignment). It may be useful to make a plan of the way that you wish to approach the task. This will ensure a logical approach is taken. Additional technical information will also be required before you start.

WWW **www.autodata.com**

Multiple choice questions

Choose the correct answer from a), b) or c) and place a tick [✓] after your answer.

1 **How is the alignment of the front wheels adjusted?**

a) By lengthening or shortening the drag link [ ]

b) By lengthening or shortening the track rod [ ]

c) By lengthening or shortening the steering column [ ]

2 **What type of oil is used in a fluid type power assisted steering system?**

a) 5W30 multigrade [ ]

b) EP 90 [ ]

c) ATF [ ]

3 **To check a vehicle wheel alignment a technician must use?**

a) Engineers rule [ ]

b) Tracking gauges [ ]

c) External micrometer [ ]

4 **What does KPI stand for?**

a) Kingpin inclination [ ]

b) Kingpin identification [ ]

c) King pivot inclination [ ]

5 **Where would a technician find the maximum amount of wear that is permitted in a track rod end?**

a) Vosa manual [ ]

b) Drivers' handbook [ ]

c) Service sheet [ ]

6 **One possible cause of the outer edge of the front tyres wearing faster than the rest of the tyre tread is?**

a) Vehicle is toeing out [ ]

b) Vehicle is toeing in [ ]

c) Overinflation [ ]

SECTION 3

Vehicle braking systems components and maintenance

USE THIS SPACE FOR LEARNER NOTES

_____

_____

_____

Learning objectives

After studying this section you should be able to:

- **Work safely on vehicle braking systems.**
- **Identify non-ABS vehicle braking system components.**
- **State how basic vehicle braking systems operate.**
- **Undertake routine maintenance on vehicle braking systems.**

Key terms

Drum brake   A brake in which curved shoes press on the inside of a metal drum to produce friction in order to slow the vehicle down.

Disc brake   A brake in which friction pads grip a rotating disc in order to slow the vehicle down.

Leading shoe   One of the shoes in a brake drum assembly which pivots outwards into the drum first.

Trailing shoe   The shoe in a brake drum assembly which is forced away from the drum by its rotation.

Self-servo action   Self-energizing effect which helps to multiply the braking force when the brake shoe contacts the drum.

Caliper   Housing for the piston(s) which moves the brake pads to contact the rotating disc.

Hygroscopic   An ability to absorb moisture from the atmosphere.

Master cylinder   The largest cylinder in the hydraulic circuit which pressurizes the fluid.

Brake fade   Loss of friction and braking force due to overheating brakes.

L	N	C	N	B	G	N	I	L	I	A	R	T	N
N	D	E	P	F	N	L	E	A	D	I	N	G	O
E	I	I	F	R	I	C	T	I	O	N	I	E	O
P	L	A	S	I	Z	I	H	F	U	I	B	D	R
R	I	S	M	C	I	P	Y	E	S	U	R	D	D
F	E	S	R	T	G	O	D	I	E	U	A	E	O
D	I	I	T	I	R	C	R	R	M	D	K	D	L
N	L	Y	E	O	E	S	A	I	O	G	E	A	D
R	R	A	G	N	N	O	U	E	E	E	P	F	B
C	A	L	I	P	E	R	L	G	O	F	A	E	D
R	D	I	U	L	F	G	I	G	A	H	D	K	R
R	E	D	N	I	L	Y	C	R	E	T	S	A	M
D	C	A	R	S	E	H	C	A	A	F	F	R	P
I	E	G	H	O	S	L	R	I	E	L	R	B	C

FRICTION

BRAKEFADE

FLUID

HYDRAULIC

MASTERCYLINDER

HYGROSCOPIC

BRAKEPADS

PISTON

CALIPER

DRUM

SHOE

SELFENERGIZING

TRAILING

LEADING

DISC

FRICTION

Remember, when working with braking systems you may be dealing with:

- Brake dust
- Hydraulic fluid under pressure
- Chemicals which are carcinogenic or toxic
- Rotating components
- Waste disposal
- Hot components

Hazardous substances

Many liquids and substances used are either:

- Toxic
- Corrosive
- Irritants
- Harmful

Ensure you are aware of the correct precautions to take when working with them.

Control of Substances Hazardous to Health (COSHH) safety data sheets should be made available for all of these substances and they will state the precautions and actions to take.

THE BRAKING SYSTEM

Brakes are all vital for safe vehicle control and comfort. Most parts of these systems are hidden from view. On a bicycle you use the brakes to slow down or stop. The brakes on a car serve the same purpose only they are operated by the footbrake which is connected to the braking units in the wheels by pipes containing hydraulic fluid.

Braking system principles

Complete this paragraph on the brake system using words from the following word bank (note: there are two extra distracter words):

pedal	**foot**	**increased**	**diameter**
force	**shoes**	**pressure**	**cylinders**
compressible	**decreased**	**transfer**	**master**

A hydraulic fluid system uses brake fluid to _____ pressure applied from the brake

_____ to the pads or _____. Brake fluid is non-_____, which means that the

pressure applied by your _____ is transmitted equally to each of the brakes. However, the

size of the piston _____ of the _____ cylinder (by your foot) and the pistons in the

wheel _____ that operate the pads or discs is different so the force is _____ to

improve and multiply the braking _____.

The main components of a typical braking system are shown below. Match the following terms to the correct component photo:

_____

master cylinder   **Disc brake**   **Drum brake**

_____

_____

WHEEL CYLINDERS

The wheel cylinder consists of a simple cylinder and piston(s) which, when the pistons are forced outwards by fluid pressure, activates the brake shoes. A double-piston type is shown below. Label the drawing using the words below. Note: some words may be used more than once or not at all:

dust cover　　　　**fluid seal**　　　　**centralizing spring**　　　**cylinder body**
piston　　　　　　**bleed nipple**　　　**circlip**　　　　　　　　**brake fluid**

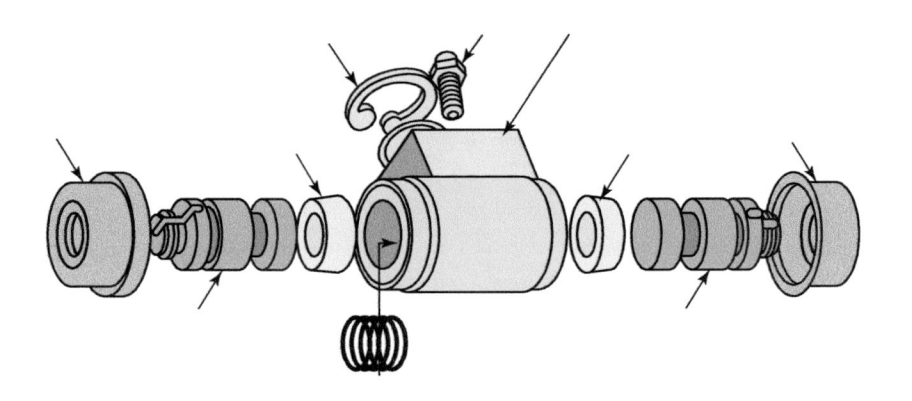

Why are the rubber seals needed on the pistons?

_____

Which operating pistons are normally larger in diameter, the master or slave cylinder?

_____

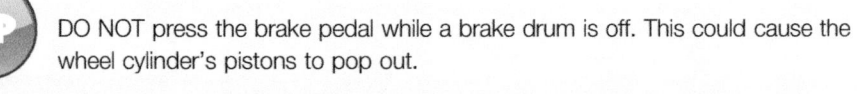

TIP DO NOT press the brake pedal while a brake drum is off. This could cause the wheel cylinder's pistons to pop out.

DRUM BRAKES

The wheel cylinder pistons and brake shoes are mounted on a back plate which is secured to the vehicle's axle. The brake drum is rotating at road speed, as the brake is pressed the shoes move outwards and rub against the rotating drum. This friction slows the vehicle.

LEADING AND TRAILING BRAKE SHOES

The figure below shows a 'double-piston wheel cylinder' acting on a _leading brake shoe_ and a _trailing brake shoe_.

Identify the parts in the drum brake assembly from the list below:

trailing shoe　　　　　　**twin piston wheel cylinder**　　**brake shoe pivot**
leading shoe　　　　　　**hold-down spring**　　　　　　　**return spring**
automatic adjuster　　**parking brake lever**

Explain the purpose of the components listed below.

Wheel cylinder: _____

Return spring: _____

Adjuster: _____

Parking brake lever: _____

_____

What friction materials are used for brake linings?

_____

What health hazards are associated with the dust from this friction material?

_____

When two surfaces rub together, what is the force known as? _____

What is generated when two surfaces are rubbed together? _____

PARKING BRAKE

On most cars, the parking brake works only on the rear wheels.

What do the simplest systems use to operate the parking brake?

_____

_____

The handbrake operation on many modern cars is now undertaken by?

_____

The main parts of a typical braking system are shown below. Label the diagram.

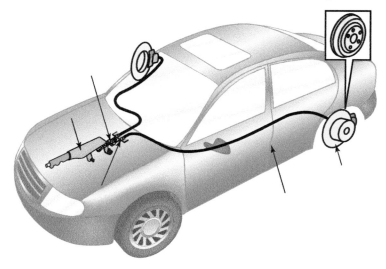

Look at the drawing of the drum brake below. Label the main parts of the handbrake system using the words below:

parking brake lever   **parking brake cable**   **secondary brake shoe**
compensator   **return springs**   **left rear brake**

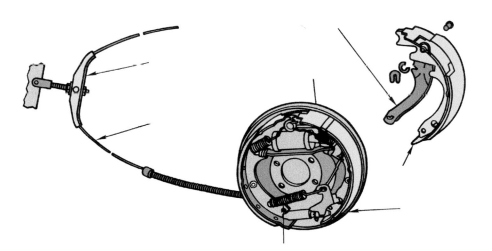

Complete the paragraph using the following words (note: there are two extra distracter words):

inspection	cables	released
levers	shoe and drum	compensator
pivot	three clicks	six clicks

When carrying out a vehicle _____, the handbrake should only travel for approximately

_____ of the ratchet before it is applied. On a typical _____ parking brake

system, when the handbrake is applied the _____ and _____ pull on the

parking brake _____. These move the brake shoes outwards against the drum until the

handbrake is _____.

Remove a rear brake drum and ask someone else to operate the handbrake. Observe how the brake shoes are moved by the mechanism.

Look under the car and watch how the cables and linkage are connected and move.

Identify, with the help of technical data, where and how to adjust the vehicle's handbrake mechanism.

Examine a vehicle braking system. Note where the brake pipes are connected to and where the discs, pads, drums and brake shoes rub against each other.

Draw and label a diagram showing the location of all of the components in a vehicle's braking system. Ensure you show in detail where the brake pipes are connected.

REMOVING AND REPLACING REAR SHOES

This task can be very tricky and needs to be undertaken in a logical manner. You will need to access manufacturers' specific information before you start.

What should a technician do before removing the brake shoes?

_____

_____

_____

_____

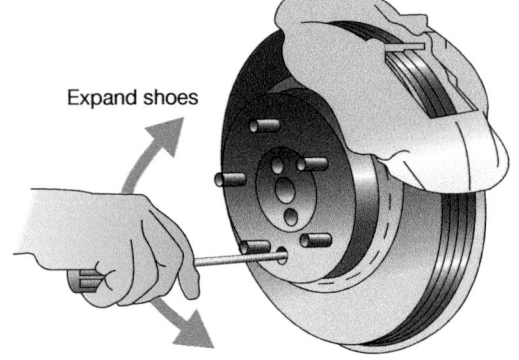

Expand shoes

Retract shoes

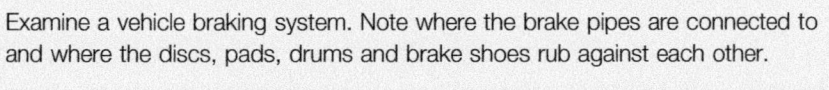

The adjusters for some auxiliary drum parking brakes are accessible through the outboard drum surface.

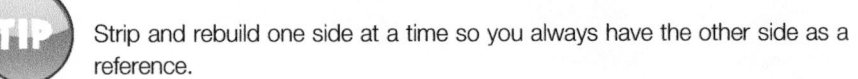

TIP

Strip and rebuild one side at a time so you always have the other side as a reference.

DISC BRAKES

Sliding or fixed caliper disc brake

With **disc brakes** the hydraulic pistons clamp the pads onto the disc rota. The braking action is similar to that of bicycle brakes. The two main types of calliper are sliding or fixed type.

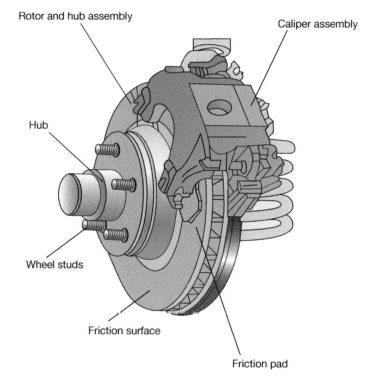

Rotor and hub assembly

Caliper assembly

Hub

Wheel studs

Friction surface

Friction pad

Basic parts of a disc brake assembly.

Sliding or floating calipers

The sliding or floating **caliper** is by far the most popular and only has one piston.

Complete the paragraph using the following words (note: there are two extra distracter words):

handbrake	out	pins	caliper frame	slides
other	slide	footbrake	applied	disc
one	fixed	direction	pressure	

The caliper is mounted on _____ that let it _____ when the brakes are

_____. When you press the _____, the increase in fluid _____ forces the

piston _____, and the _____ in the opposite _____. The frame 'floats' or

_____. The piston applies _____ pad and the frame applies the _____

pad. Equal force is then applied to both pads which clamp against the _____.

Fixed calipers can have two, three or four pistons. They have pistons located on each side of the disc. The housing of the caliper does not move during braking.

In small groups investigate the advantages and disadvantages of both types of disc brake caliper.

Study the drawing of the disc brake caliper shown below and name the main parts.

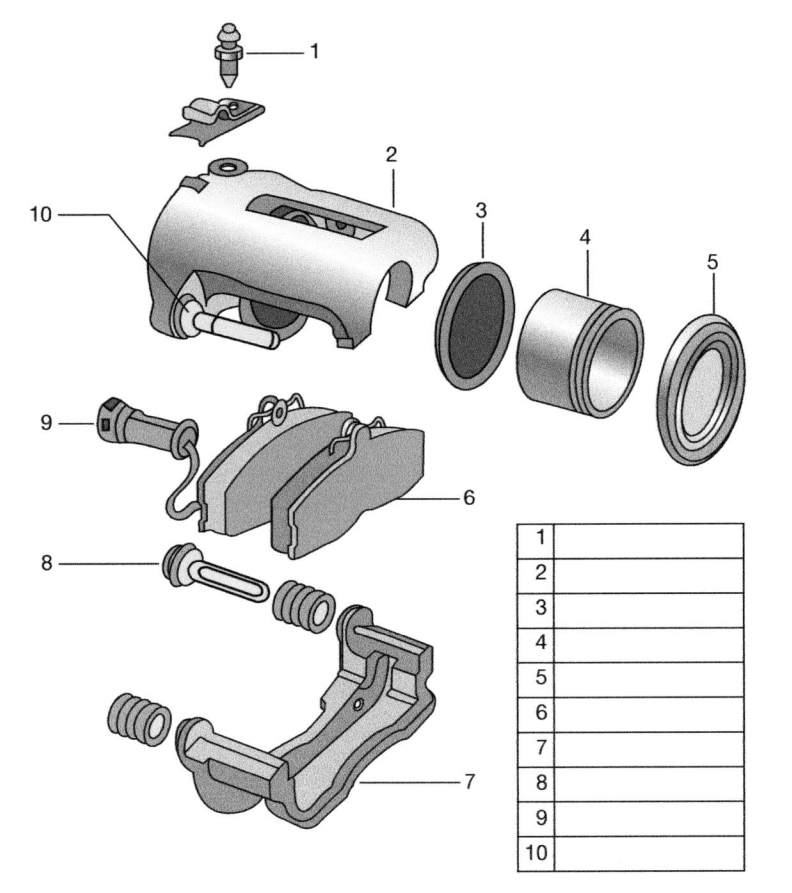

1	
2	
3	
4	
5	
6	
7	
8	
9	
10	

On the disc brake there is a cable attached to one of the pads. What is it for?

_____

_____

RMP The sliding frame of the caliper may become tight, or seize up. When you renew the brake pads, always check that the frame moves freely. Many modern cars have disc brakes fitted to all four wheels.

The drawing below is of a brake master cylinder vacuum servo.

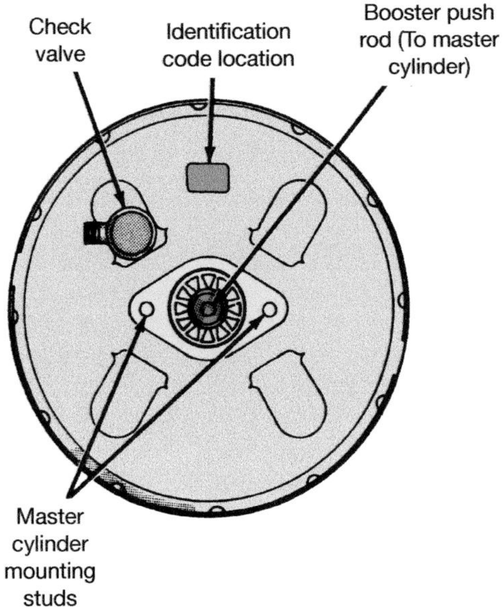

Check valve

Identification code location

Booster push rod (To master cylinder)

Master cylinder mounting studs

Front view

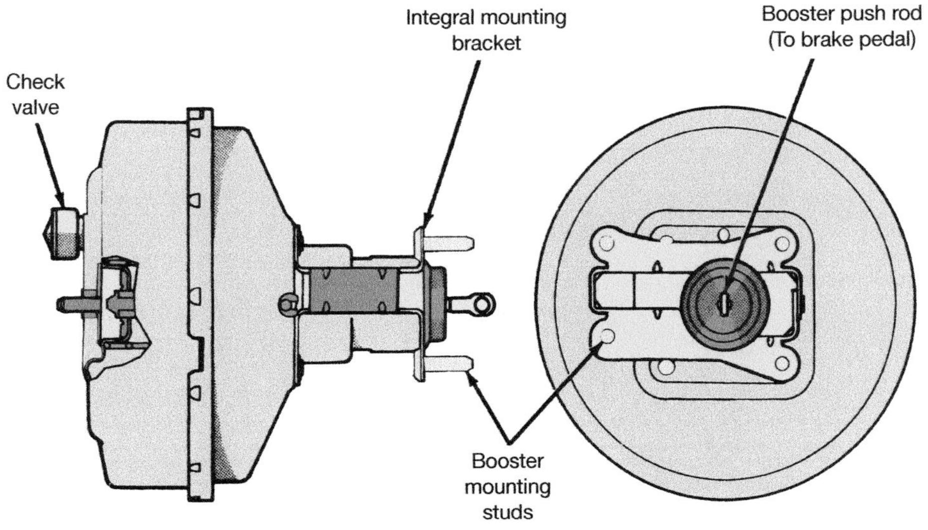

Integral mounting bracket

Booster push rod (To brake pedal)

Check valve

Booster mounting studs

Side view

Rear view

What does the brake servo do?

_____

 To keep the braking system in good working order, you will need to inspect, test, adjust, repair and replace parts of the system.

Before removing and replacing brake pads and brake shoes on a vehicle it is important to adhere to the correct procedures. Make a list of what should be included:

- _____
- _____
- _____
- _____
- _____
- _____

BRAKE ADJUSTMENT

It is important to keep the **brake shoe** or **brake pad** material in the right position – just clear of the drum or disc.

What would happen if the brake adjustment was incorrect?

_____

_____

Most **brake adjusters** work automatically. Some vehicles, though, have *manual* adjusters at the rear.

What would have to be undertaken on a vehicle fitted with manual adjusters?

_____

REMOVING AND REPLACING BRAKE PADS

The images below show a technician removing and refitting a set of front brake pads. Number the images in the correct sequence to match the correct descriptions of the procedure on page 156.

_____

_____

_____

_____

1   Front brake pad replacement begins with removing brake fluid from the master cylinder reservoir.
2   Raise the car. Make sure it is safely positioned on the lift. Remove its wheel assemblies.
3   Inspect the brake assembly. Look for signs of fluid leaks, broken or cracked lines, or a damaged brake rotor. If any problem is found, correct it before installing the new brake pads.
4   Loosen the bolts and remove the pad locator pins.
5   Lift and rotate the caliper assembly from the rotor.
6   Remove the brake pads from the caliper assembly.
7   Fasten a piece of wire to the car's frame and support the caliper with the wire.
8   Check the condition of the locating pin insulators and sleeves.
9   Place a piece of wood over the caliper's piston and install a C-clamp over the wood and caliper. Tighten the clamp to force the piston back into its bore.
10 Remove the clamp and install new locating pin insulators and sleeves, if necessary.
11 Install the new pads into the caliper.
12 Set caliper with pads over the rotor and install the locating pins. After the assembly is in the proper position, torque the pins according to specifications.

BRAKE BLEEDING

The process of forcing the air out is called **bleeding.** There are special bleed nipples in the calipers and the wheel cylinders. These can be released, allowing air to be pumped out of the system.

MANUALLY BLEED THE BRAKE SYSTEM

Complete the paragraph using the following words (note: there are two extra distracter words):

hose	close	air bubbles	master
release	brake	slowly	
wheel	piston	furthest	

Open the bleeder screw that is the _____ from the _____ cylinder and ask a workmate to

_____ push the _____ pedal down. With a clear bleeder _____ inserted into a jar, you can see the

_____ passing down the tube. _____ the bleeder nipple off, and have the workmate slowly

_____ the brake pedal.

Repeat this process until there are no more air bubbles coming out of the brake cylinder and the new brake fluid can be seen.

What else must now be undertaken?

_____

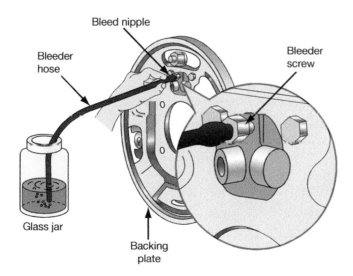

Bleed nipple

Bleeder hose

Bleeder screw

Glass jar

Backing plate

TIP Bleed nipples must be tightened to the manufacturer's specifications; they can easily be snapped, particularly when **undoing** them.

What checks should be undertaken after replacing flexible brake hoses?

● _____

● _____

_____

On the diagram below indicate, by numbering 1 to 4, which cylinder has to be bled first and which will be the last to be bled.

How could air get into the hydraulic system?

_____

_____

157

What would be the effects of air in the hydraulic system?

- _____
- _____

Brake fluid has a hygroscopic nature, what is meant by this?

_____

What effect will moisture have on brake fluid if absorbed?

_____

There are many different types of drum and disc brakes. Calipers, wheel cylinders, adjusters, handbrake systems and so on may be quite different for each of these. As you work under supervision on a variety of vehicles, you will gradually build the skill and knowledge to do this work safely.

 Now conduct a brake test and record your findings below. Record five readings for each test:

Brake test	Handbrake	Footbrake	Balance
1			
2			
3			
4			
5			

Multiple choice questions

Choose the correct answer from a), b) or c) and place a tick [✓] after your answer.

1 **When you release the brakes, which component draws the brake shoes away from the drums?**

 a) Return springs [ ]

 b) Shoe retainers [ ]

 c) Handbrake mechanism [ ]

2 **Which part locks the handbrake on?**

 a) Pawl [ ]

 b) Ratchet [ ]

 c) Return spring [ ]

3 **Name one way of checking brake efficiency:**

 a) Bleeding the brakes [ ]

 b) Brake roller testing the brakes [ ]

 c) Visually inspecting the brakes [ ]

4 **What is the legal minimum efficiency for foot and handbrake operation?**

 a) footbrake: .... %

 b) handbrake: .... %

5 **What would happen if one hydraulic pipe leaked?**

 a) All brakes would fail [ ]

 b) Two brakes would fail [ ]

 c) Handbrakes would fail [ ]

6 **Name the braking system that prevents wheel lock or skidding.**

 a) SRS [ ]

 b) ASR [ ]

 c) ABS [ ]

Vehicle wheels and tyres construction and maintenance

USE THIS SPACE FOR LEARNER NOTES

_____

_____

_____

_____

Learning objectives

After studying this section you should be able to:

- Work safely when working with road wheels and tyres.
- Identify how wheels and tyres are constructed.
- Understand wheel and tyre terminology.
- Undertake routine maintenance and replacement of road wheels and tyres.

Key terms

Radial A tyre in which the plies are placed at right angles to the rim.

Cross ply A tyre in which the plies are placed diagonally across each other at an angle of approximately 30 to 40 degrees.

Well base rim A rim with a centre channel which enables easy removal and refitting of the tyre.

Aspect ratio The ratio between the height and width of a tyre (expressed as a percentage).

Tube Fitted inside the tyre to retain air.

Plies Increase the strength of the tyre.

Load index How much weight a tyre will safely carry.

Speed rating Maximum safe tyre speed.

Balancing Correcting any unbalance of wheel and tyre assemblies.

When working with wheels and tyres you may be dealing with:

- Jacking and supporting vehicles
- High air pressures and pressurized components
- Rotating machinery
- Chemicals and solvents
- Waste disposal

```
L R P C S A P D E S O      RADIAL
L T P N P O O S P I L      BALANCING
Y L I A S E B U T O L      SPEEDRATING
A A L S I T N A A A E      LOADINDEX
E B R N E P R D I N W      PLIES
L L A T L T I D C R D      TUBE
I G N I C N A L A B P      ASPECTRATIO
S P E E D R A T I N G      WELL
B S P E E G L P C N O      CROSSPLY
L S X L D E A E D P C
A C R O S S P L Y B S
```

WHEELS AND TYRES

When accelerating, braking and steering, the road wheel assembly (the tyre and metal wheel) has to transmit the drive safely to the road surface. It also provides a small amount of softness to absorb any bumps in the road.

What could happen if the driver fails to check his/her wheels and tyres are in good condition?

_____

 **TIP** The tyre is all that keeps a vehicle in contact with the road surface.

Name FIVE factors that determine the type of tyre fitted to a vehicle:

1 _____

2 _____

3 _____

4 _____

5 _____

TYRE TREAD PATTERN DESIGN

The design of a tyre, particularly the tread pattern, is dictated by the type of vehicle it is fitted to and how the vehicle will be used.

State the name of and a typical application for the three different tyres shown below.

Bridgestone

_____   _____   _____

_____   _____   _____

A tread pattern is formed into the tyre during the manufacturing process. The tyre is placed in a mould and heated. The tread pattern is formed in the casing, and the tyre is then allowed to cool.

For more information on tyre manufacture visit **http://www.kwik-fit.com/ tyre-manufacturers.asp**

Why is a tread pattern required on a tyre?

_____

_____

What would happen if the tyre did not remove all of the water?

_____

In the figure below small bars or bumps (properly known as tread wear indicators) can be seen at the bottom of the tread pattern.

Tread wear indicators

Tread wear indicators

Bridgestone

Why do the tyre manufacturers build small lumps into the tread? _____

_____

Which tyre has the most grip on a dry surface, a tyre with tread or a tyre with no tread? Explain your choice.

_____

_____

TYRE CONSTRUCTION

The tyre is a flexible rubber casing which is mainly synthetic. It is reinforced or supported by other materials, for example, rayon, cotton, nylon and steel.

What effect will adding more plies or reinforcing have on a tyre?

_____

Types of tyre construction

The two principal types of tyre construction used on road vehicles are:

1 _____

2 _____

Most modern vehicles are fitted with _____ tyres. A tyre casing consists of plies which are layers of different material looped around each bead to form a case or carcass.

What is the basic difference in structure between the two types of tyres you listed above?

_____

With _____ tyres, the construction of the ply 'cords' is diagonal from the angle of the tyre bead. This is approximately _____ degrees.

_____ cords form an angle of _____ degrees to the tyre bead.

Indicate on the two diagrams below the direction of the cords in relation to the tyre bead for the two types of tyre construction.

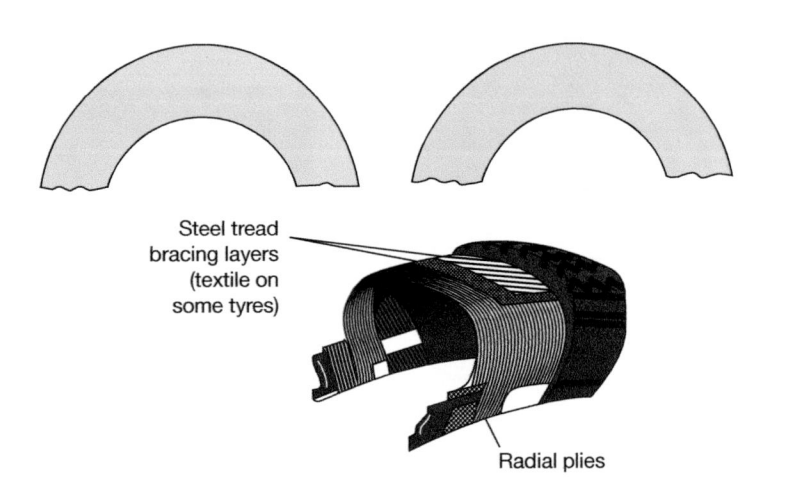

Steel tread bracing layers (textile on some tyres)

Radial plies

Tubeless tyre

In the tubeless tyre, the casing cords run *radially* from bead to bead. The middle section of the tyre is the 'belt'. It may have more than one layer of textile or steel cords (or both) beneath the tread.

What are the advantages of more bracing layers?

_____

_____

Tubeless tyres do not require an inner tube. This type of tyre is made with a soft lining on the inside making the complete wheel and tyre assembly airtight.

Name one of the main advantages of a tubeless tyre:

_____

_____

_____

Cross plies

Tread

Apex strip

Side wall

Inner lining

Bead wires

Bead

Textile bracing layers

Bias (diagonal) plies

Tubed tyre

Complete the paragraph using the words below (note: there are two extra distracter words):

Inside	expands	inner	bead
rim	casing	washers	angle
tyre	inflated	airtight	

This type of tyre requires an _____ tube to seal the air _____ the tyre. The casing cords run at an _____ from bead to _____. A special, _____ valve assembly is needed. This fits into the _____, it can be held with a nut and sealing _____. The tube is _____ inside the tyre casing and as it _____ the tyre is held onto the wheel _____, just like a bike tyre.

What is the major disadvantage of using an inner tubed tyre?

_____

_____

What special precautions need to be taken when removing or refitting a tubed tyre?

_____

The images below show the two different designs of tyre valves, tubed and tubeless.

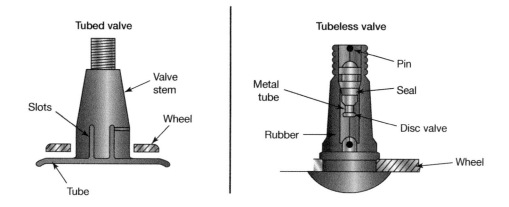

What is the purpose of the slots in the outer stem of the tubed valve?

_____

What type of valve core is used in both valves shown?

_____

 For information on the parts of a tyre and much more, visit:
www.carbibles.com/tyre

Below is a sectioned view of a tyre. Label the different sections of the tyre listed below.

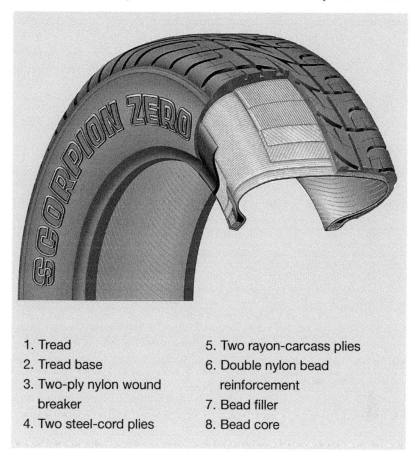

1. Tread
2. Tread base
3. Two-ply nylon wound breaker
4. Two steel-cord plies
5. Two rayon-carcass plies
6. Double nylon bead reinforcement
7. Bead filler
8. Bead core

Tyre wall markings

When looking at a tyre, most people will notice the name of the manufacturer because it is in big letters. As well as the manufacturer's name, the rim size, the type of tyre construction, aspect ratio, maximum load and speed are all displayed on the side wall.

Look at the drawing below.

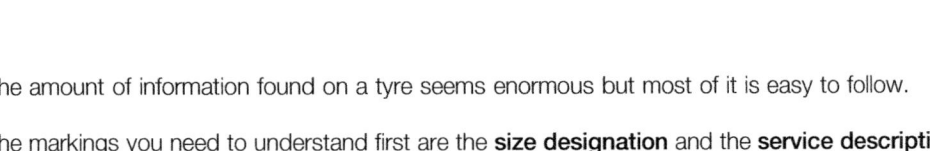

- Brand
- Speed rating
- Load index
- Diameter of wheel in inches
- Radial
- Aspect ratio (ratio of width to height)
- Width of tyre in millimetres
- P = Passenger car tire
- Model
- Max. inflation
- Tyre ply composition and materials used
- DOT size code
- DOT type code
- DOT data code
- Max. load
- Treadwear
- Traction
- Temperature
- UTQG ratings

The amount of information found on a tyre seems enormous but most of it is easy to follow.

The markings you need to understand first are the **size designation** and the **service description**. Together these indicate the tyre's *dimensions* (its size), *structure, load capacity* and *speed rating*.

TIP Incorrectly sized wheels and tyres can overload wheel bearings and change steering characteristics. Using different sized wheels and tyres is illegal.

Select a vehicle and write down the important information displayed on the side wall of the tyre.

_____

_____

_____

_____

Let's say a tyre has a size displayed as 185 × 60 R 14 82 H. What do the numbers and letters tell us?

Symbol	Meaning
185	
60	
R	
14	
82	
H	

WWW For more detail of what these markings mean, visit:
http://www.pirelli.com/tyre/gb/en/car/genericPage/all_about_tyres

Tyre aspect ratio

This ratio shows the relationship between the section *height* (of the side wall) and the section *width* (the width of the tread). The ratio is expressed as a percentage. The normal or **standard aspect ratio** for radial-ply tyres is 82%.

Standard aspect ratio   60% aspect ratio

50% aspect ratio

If a tyre has a width of 200 mm and has a 50% aspect ratio, how high would the tyre wall

measure in millimetres? _____

If a tyre is displaying 385 × 65 × 22.5 on its side wall what does the 22.5 indicate?

_____

Tyre marking

The letter on the side wall tells you whether the tyre is of cross-ply or radial-ply construction.

Most modern tyres are _____. They display the letter '_____' or the word '_____'.

Tyre speed rating symbol

This letter tells you the _____ for any particular tyre. Usually it is shown at the end of the size designation.

Tyre load index

The **load index** number states the _____ that *one tyre* can safely carry. You can check this on a **load index (LI) table**.

A part of a speed symbol table is shown below. These are the most common symbols.

Part of a load index table

LI	kg	LI	kg	LI	kg	LI	kg	LI	kg	LI	kg
0	45	20	80	40	140	60	250	80	450	100	800
1	46.2	21	82.5	41	145	61	257	81	462	101	825
2	47.5	22	85	42	150	62	265	82	475	102	850
3	48.7	23	87.5	43	155	63	272	83	487	103	875

Tyre speed symbol markings table

Symbol	km/h	mph
Q	160	100
R	170	105
S	180	113
T	190	118
H	210	130
V	240	149
VR	210+	130+
Y	300	186
ZR	240+	150+

Locate three very different vehicles. List the makes and models and all of the tyre information displayed on the side wall

	Make and model of car		
Tyre information	*1*	*2*	*3*
Product name			
Size designation			
Service description			
Tyre size			
Manufacturers' name			
ECE tyre approval mark/number			
Load capacity			
Speed rating			

Study your findings and from the speed and load tables provided, identify:

The tyre with the highest speed rating.

_____

The tyre with the largest load carrying capabilities.

_____

Now, in small groups compare your findings.

LIGHT VEHICLE WHEELS

There are many small differences between wheels. These are very important. Always check the *rims,* which must be of the correct size and type.

If the correct size of tyre is fitted, but to the wrong type of rim, this could cause an accident.

What forces must a wheel rim be strong enough to withstand?

_____

_____

What are the advantages of alloy wheels compared with pressed steel rims?

_____

_____

_____

MODERN RIMS

Below is a selection of modern rim profiles, suitable for tubeless radial tyres.

Notice the **humps** close to the outer edges of the rims.

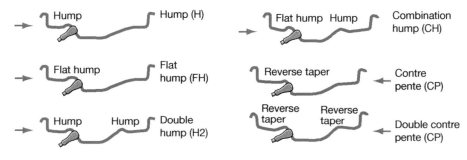

Why are these humps so important? _____

_____

TIP

There are many different types of wheels including pressed steel, space saver and split rim. It is important to remove and to fit tyres from the correct side of the rim.

If you remove or fit the tyre from the *wrong* side, you might damage the tyre, and you might injure yourself.

Name the part of a rim that allows the tyre to be removed easily:

_____

Complete the paragraph with words from the following word bank (note: there are two extra distracter words):

wheel	inflated	airtight	upwards
15	pressure	ridges	rim
90	seats	grip	drop-centre

The wheel rims shown are referred to as _____ rims; this is because the bead _____ are inclined _____ from the centre of the _____ at an angle of approximately _____ degrees. This slow taper gives the tyre a good _____ onto the rim, and forms an _____ seal. When the tyre is correctly _____, it is locked to the rim by air _____ and the angle of the bead or by safety _____ or humps, close to the flange.

The well allows one side of the tyre to drop into it during fitting or removal of the tyre. Indicate on the diagram below where the well is located.

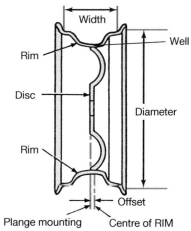

State THREE effects of the tyre pressures being too high and three effects of tyre pressures being too low.

Pressures too high:

1 _____

2 _____

3 _____

Pressures too low:

1 _____

2 _____

3 _____

 **TIP** Some special high-performance tyres need special high-performance rims. It is vital that the tyres and rims match together exactly.

WHEEL RIM CONDITION

For safety, wheel rims must be checked to ensure they are in a serviceable condition.

Apart from obvious impact damage, what TWO other things should you look for?

1 _____

2 _____

What could happen if rust forms in the bead area of the wheel rim?

_____

_____

TIP Sometimes fine cracks are very hard to see. Cleaning the rim thoroughly will help you spot them.

WHEEL SECURITY

Wheels are fastened to the hubs by wheel studs, nuts or bolts.

The threads between the studs and nuts are close fitting and accurately sized.

All wheel nuts must be tightened in the correct sequence otherwise the wheel may not sit correctly on the hub.

Show the correct sequence for tightening wheel nuts on the drawing below.

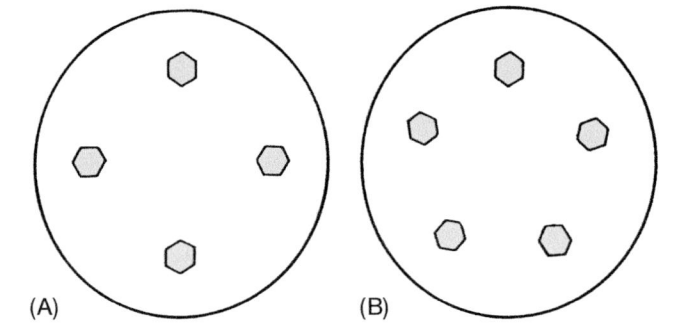

(A)   (B)

Securing a wheel to a vehicle is one of the most important jobs you will do as a tyre technician. **If you do not fix the wheel properly you could cause an accident or even death!**

Securing a wheel is not simply a matter of tightening up the wheel nuts and bolts. It is not a test of your muscle power. For a wheel to be properly secure, TWO things must happen. Complete the following points:

1 The wheel must be _____ on the hub.

2 The wheel nuts and bolts must be tightened to the _____.

What specialist workshop tool shown is used to secure the wheel nuts correctly? _____

On most cars and light commercial vehicles, locating the wheel is quite simple. There is a _____ on the wheel nuts, and this matches a taper on the wheel. As the nuts are tightened, the tapers position the wheel correctly.

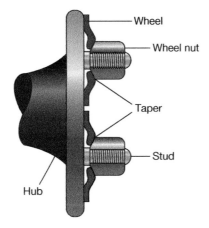

- Wheel
- Wheel nut
- Taper
- Stud
- Hub

What would happen if the wheel nut shown in the drawing was fitted incorrectly?

_____

_____

Name the type of wheel fixing shown in the image.

_____

Look carefully at wheel nuts from several vehicles. Inspect them and record your findings.

Make/model	Nut condition	Stud condition	Serviceable Yes/No

TIP Do not mix different types of wheel nuts, even if they fit onto the studs – the wheels might come loose.

Complete the following paragraph:

The *turning force* applied to a nut is called the _____. You can find out the correct torque for each nut from a workshop manual or computer database. In the UK, torque is usually shown in units of _____. The international unit of torque, though, is the _____, and this is now widely used.

What checks must be undertaken before refitting a wheel?

_____

_____

In small groups choose *three* very different vehicles. Find the wheel nut torque setting and then fill in the table below.

Make	Vehicle type	Model	Type of fixing	Wheel nut torque lb/ft or Nm

TIP After you have tightened the wheel nuts – especially on a commercial vehicle – advise the driver to check the tightness (but not *over*-tighten) from time to time. A common recommendation is that the first check should be between 50 km and 250 km.

REMOVE AND REFIT A TUBELESS TYRE

Workshop task

Below are a series of images of a technician removing and refitting a tyre. Study the images and place them in the correct order for the task using the descriptions of the correct procedure on page 171.

1 Dismounting the tyre from the wheel begins with releasing the air, removing the valve stem core, and unseating the tyre from its rim. The machine does the unseating. The technician merely guides the operating lever. Once both sides of the tyre are unseated, place the tyre and wheel onto the machine. Then depress the pedal that clamps the wheel to the tyre machine.

2 Lower the machine's arm into position on the tyre and wheel assembly.

3 Insert the tyre iron between the upper bead of the tyre and the wheel. Depress the pedal that causes the wheel to rotate. Do the same with the lower bead.

4 After the tyre is totally free from the rim, remove the tyre.

5 Prepare the wheel for the mounting of the tyre by using a wire brush to remove all dirt and rust from the sealing surface. Apply rubber compound to the bead area of the tyre.

6 Place the tyre onto the wheel and lower the arm into place. As the machine rotates the wheel, the arm will force the tyre over the rim. After the tyre is completely over the rim, install the air ring over the tyre. Activate it to seat the tyre against the wheel.

7 Reinstall the valve stem core and inflate the tyre to the recommended inflation.

Once you feel confident about this process, under supervision correctly remove a wheel and tyre from a vehicle and practise this skill.

What information will a technician require **before** replacing a wheel and tyre?

___

WHEEL BALANCE

When customers complain of **vibration** felt through the steering wheel, usually at about 45–50 mph, the reason is probably that the wheel is **out of balance**. If it is really bad, the whole car may shake.

You probably already know more about wheel balance than you realize. For example, suppose that you lifted the front of a bicycle and allowed the wheel to spin freely. Where would the valve be when the wheel stopped? Mark the valve's position on the drawing below.

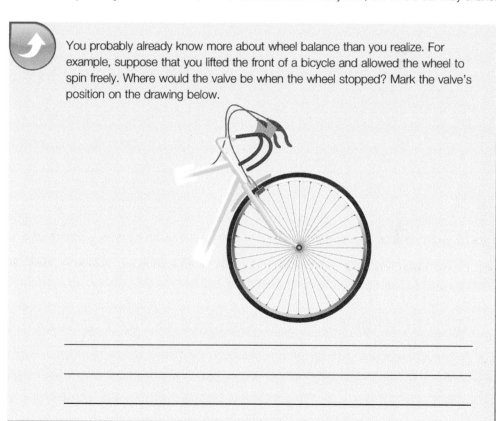

___

___

___

If you rotated the wheel quickly, the whole bicycle would bounce up and down! You can imagine how much worse this would be with a car wheel at speed.

Where would you wind some wire (about the same weight as the valve) on the wheel to enable it to be balanced?

_____

Dynamic balance

Motor vehicle tyres are much wider than those on bicycles. Therefore static balance on its own will not cure *all* wheel balance problems. The wheels also need **dynamic balance**. This is undertaken on a specially designed machine called a 'wheel balancing machine' shown opposite.

Tyre/wheel balancing equipment

There are many types and makes of **wheel balancing machines**. All wheel balancing machines will:

1 _____

2 _____

3 _____

When is it necessary to balance tyre and wheel assemblies?

1 _____

2 _____.

3 _____

4 _____

5 _____

You will need to be shown how to use a wheel balancer. For your own safety, follow the directions carefully. When the wheel spins, dirt or grit may fly out towards your face and eyes. Keep to one side to avoid it and wear safety goggles.

When a vehicle is in a workshop, a trained technician will always visually check the tyres. If it is in for a service or repair work, always check tyres thoroughly.

WWW If you need any further guidance, visit
http://www.etyres.co.uk/uk-tyre-law

Multiple choice questions

Choose the correct answer from a), b) or c) and place a tick [✓] after your answer.

1   **As part of a service the technician would check the condition of the tyres and what else?**

a) Tyre pressures [ ]

b) Balance of the drive shaft [ ]

c) Trim height [ ]

2   **Unbalanced front wheels would cause premature wear of?**

a) Front brake pads [ ]

b) Steering and suspension components [ ]

c) Rear tyres [ ]

3   **What is the minimum tread depth for a car?**

a) 1.6mm [ ]

b) 1.8mm [ ]

c) 2.0mm [ ]

4   **What is the minimum tread depth for HGV?**

a) 1.0mm [ ]

b) 1.2mm [ ]

c) 1.4mm [ ]

5   **Nominal width of a tyre is measured in?**

a) mm [ ]

b) cm [ ]

c) m [ ]

6   **Before a tyre is removed from its rim for a puncture repair a technician should check?**

a) Sharp edges and remove balance weights [ ]

b) The correct valve stem is fitted [ ]

c) The tyre is inflated to the correct pressure [ ]

7   **One cause of the front tyres wearing the inside edge excessively would be?**

a) Excessive speed [ ]

b) Tracking [ ]

c) Balancing [ ]

PART 4
DRIVELINE TRANSMISSIONS

USE THIS SPACE FOR LEARNER NOTES

SECTION 1
Vehicle driveline maintenance 175

SECTION 1

Vehicle driveline maintenance

USE THIS SPACE FOR LEARNER NOTES

_____

_____

_____

_____

_____

_____

Learning objectives

After studying this section you should be able to:

- Identify front and rear-wheel drive layouts.
- Identify clutch components.
- Describe basic clutch operation.
- Identify gearbox components.
- Identify torque converter components.
- State the purpose of the final drive.
- Explain driveline components.

Key terms

Torque   Turning force measured in Newton metres (Nm).

Clutch plate   Component splined to the gearbox input shaft (also known as centre plate or driven plate).

Clutch slip   Sometimes referred to as feathering the clutch. When the clutch plate slips against the flywheel when the driver applies and releases the clutch pedal.

Gearbox input shaft   First shaft to turn in the gearbox, connected to the clutch plate.

Gearbox output shaft   Takes the drive to the final drive assembly.

Flywheel   Bolted to the crankshaft, transmits engine torque to the clutch plate.

Constant velocity joint   Driveshaft joint that allows steering and suspension movement when transmitting drive to the wheels.

Universal joint   Propeller shaft joint that allows small angular changes.

RWD   Rear-wheel drive.

FWD   Front-wheel drive.

AWD   All-wheel drive.

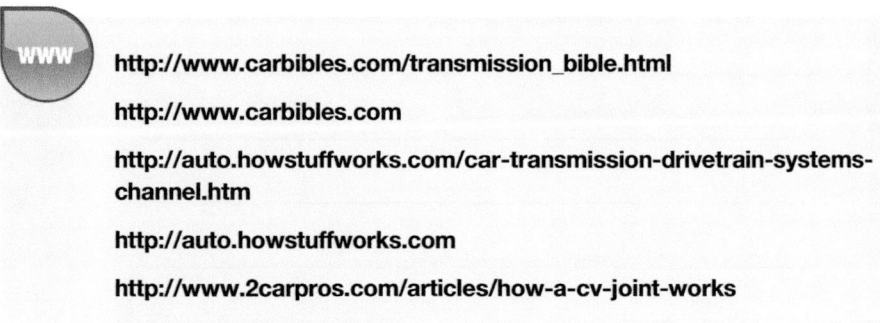
Front engine front-wheel drive

Front engine front-wheel drive (FWD) is the most popular light vehicle layout with the engine fitted sideways, this is known as **transverse.**

When the gearbox and final drive work effectively as one unit and are usually fitted transversely, the layout is known as **transaxle.**

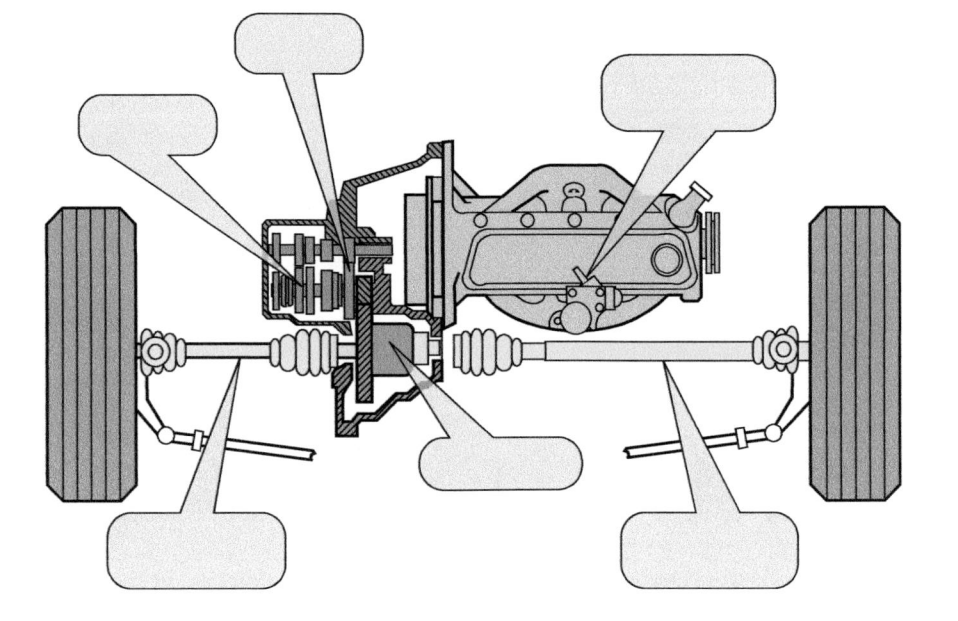

TRANSMISSION DRIVE LAYOUTS

Front engine rear-wheel drive

Complete the labels on the rear-wheel drive (RWD) diagram below. The first letter of each word is included for you:

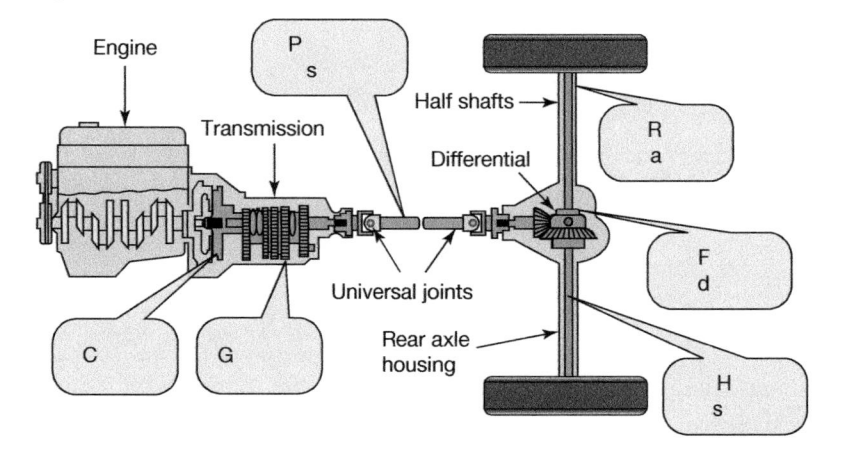

Engine

P
s

Transmission

Half shafts

Differential

R
a

Universal joints

C   G

Rear axle housing

F
d

H
s

THE CLUTCH

The clutch is part of the transmission system. It is a coupling which transmits the drive from the engine to the gearbox.

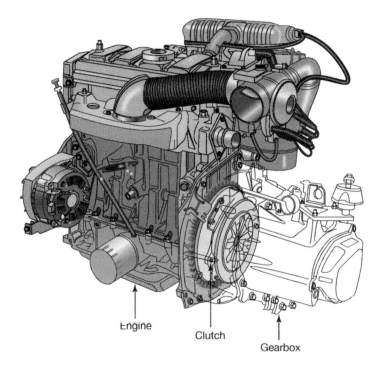

Engine   Clutch   Gearbox

The driver controls the clutch with a **clutch pedal**.

Pressing the pedal releases or disengages the clutch.

In this way the drive from the engine is connected to, or disconnected from, the gearbox.

The three functions of a clutch are:

1 **To enable the vehicle to take off smoothly from rest**
2 **To assist in gear changing**
3 **To provide a temporary position of neutral**

Clutch assembly

Label the components indicated by the arrows in the diagram below using the following terms:

release bearing	clutch plate	clutch (bell)	pressure plate
flywheel	gearbox input shaft	housing	

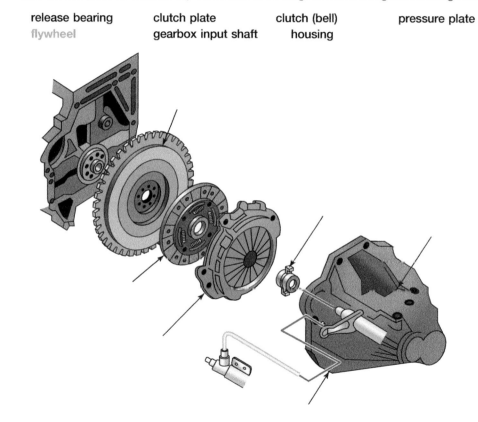

Clutch operation

This sectioned drawing shows the parts that make up the whole clutch assembly. The driver controls the clutch via the clutch pedal. Pressing the clutch pedal releases or disengages the clutch. Releasing the pedal engages the clutch. In this way the drive from the engine is connected to and disconnected from the gearbox.

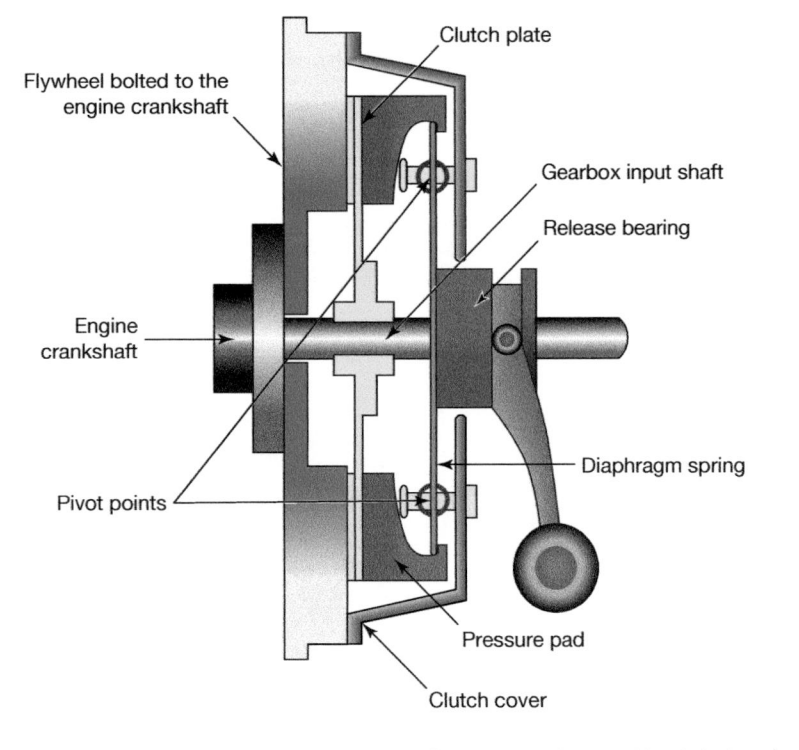

Complete the statement relating to clutch operation. Select from the word bank below (note: not all words are used):

flywheel   **diaphragm**   **clamped**   **fixed**
splined   **turning**   **stationary**

With the clutch pedal released the clutch plate is _____ between the pressure pad and the

_____ by the diaphragm spring. The clutch plate is _____ to the gearbox input shaft,

connecting the components which causes them to turn together.

This simplified drawing is showing the operation when the engine is running with the driver not pressing the clutch pedal.

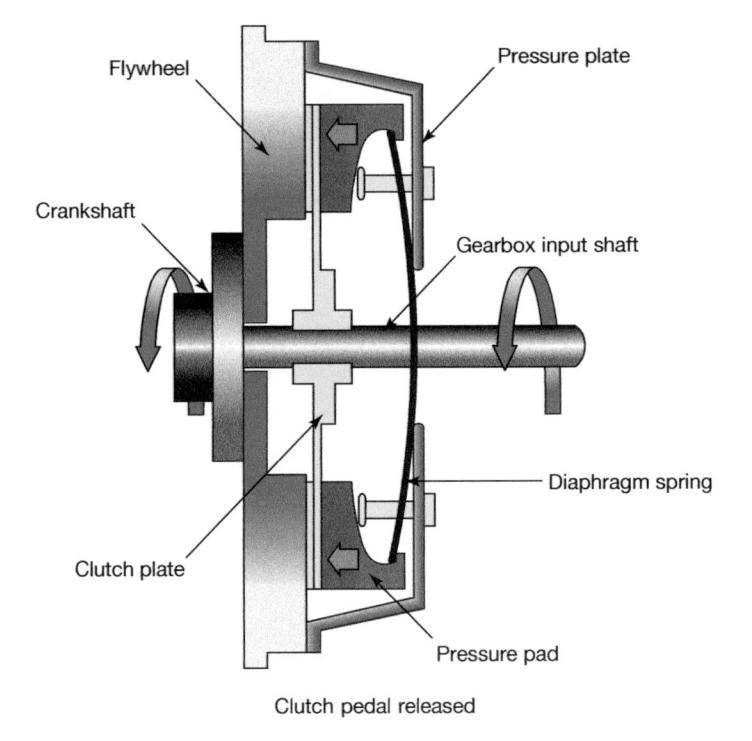

Clutch pedal released

Which of the following components will be turning? (*Tick box*)

☐ Crankshaft

☐ Gearbox input shaft

☐ Flywheel

☐ Pressure plate (includes pressure pad and diaphragm spring)

☐ Clutch plate

This drawing is showing the operation when the engine is running with the driver pressing the clutch pedal to the floor.

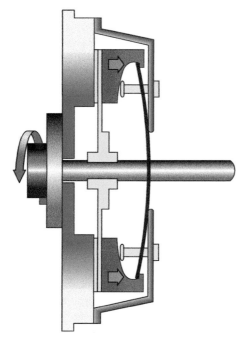

Clutch pedal pressed to the floor

Which of the following components will be turning? (*Tick box*)

☐ Crankshaft

☐ Gearbox input shaft

☐ Flywheel

☐ Pressure plate (includes pressure pad and diaphragm spring)

☐ Clutch plate

Complete the statement relating to clutch operation. Select from the word bank below (note: not all words are used):

flywheel	diaphragm	clamped	fixed
splined	turning	stationary	

When the clutch pedal is pressed the pressure pad moves away from the clutch plate as the

_____ spring is pressed and straightened. The clutch plate is _____ while the flywheel

and pressure plate revolve around it.

Clutch removal and replacement

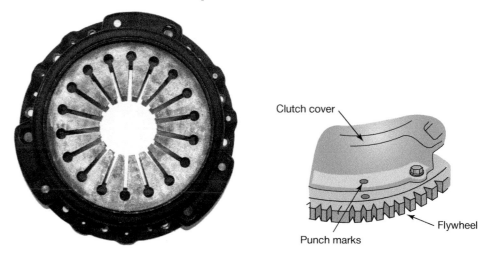

To remove the clutch plate:

1 **Slacken the clutch pressure plate bolts evenly, until they are finger tight. Wedge the flywheel to prevent it turning as the bolts are slackened.**

2 **While supporting the weight of the pressure plate, remove all of the bolts. Be careful of the centre plate to ensure it does not fall out as the bolts are removed.**

3 **Remove the pressure plate and clutch plate.**

What could happen if the bolts are not removed evenly?

_____

Clutch inspection and replacement procedure

1 **Check flywheel and pressure plate surfaces for signs of scoring, cracks or discoloration.**

Flywheel

Pressure plate

 Shade in the areas of the flywheel and pressure plate images above to show where the surfaces need to be checked.

2 **Check clutch plate for wear or damage.**
3 **Check the damper springs are not loose or broken.**

State TWO faults on the lining that could cause the clutch to slip.

1 _____

2 _____

4 **Offer the clutch plate and pressure plate into position and start the retaining bolts.**

 **TIP** Check the clutch plate positioning. It may be marked on one face with flywheel side. Alternatively, offer up clutch plate, ensuring the clutch linings touch the flywheel face without the centre hub touching the flywheel or crankshaft.

5 **Insert the clutch aligning tool and evenly tighten the bolts, while ensuring the aligning tool is free to move.**

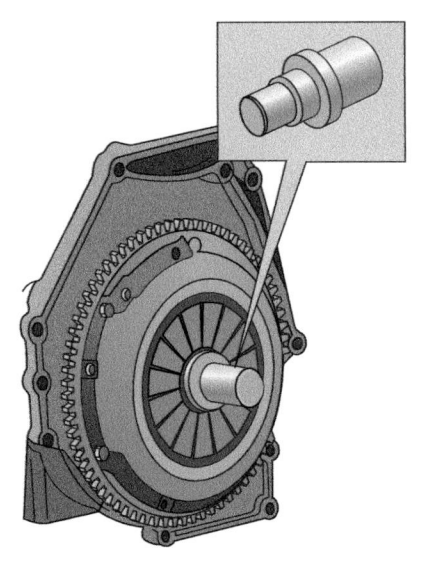

TIP Clutch aligning tools can be purpose-made for a particular vehicle or may be universal, where a selection of sizes is required.

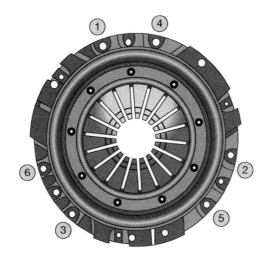

What do the numbers indicate on this drawing?

_____

_____

6 Torque **the bolts to the correct setting and remove the aligning tool.**

 TIP Torque settings can be found using workshop manuals or technical data sources (Autodata).

 Arrange with your trainer to remove and refit a clutch on a removed engine. Examine and report on the condition and refit.

Complete the table below with your findings

VEHICLE MAKE:		VEHICLE MODEL:	
Item checked	*Checking method*	*Serviceable Yes/No*	*Likely symptom if unserviceable*
Condition of the clutch release bearing	Spin the bearing and feel for roughness	Yes/No	_____ _____
Clutch plate for wear	Visual check for linings worn down to the rivets	Yes/No	_____
Clutch plate for contamination	_____ _____ _____ _____	Yes/No	Clutch slip or judder
Clutch damper springs broken or missing	Visually feel for excessive movement	Yes/No	Fierceness on take up of drive or _____
Flywheel	_____ _____ _____ _____	Yes/No	Clutch judder
Pressure plate	Visually check pad surface for scores and heat cracks	Yes/No	_____
Diaphragm spring fingers	Visually check for signs of wear on release bearing contact area	Yes/No	_____ _____ _____

CLUTCH RELEASE METHODS

Cable release

Movement of the clutch pedal is transmitted by a steel wire cable not applicable to the clutch arm moving the clutch release bearing, which presses on the diaphragm spring.

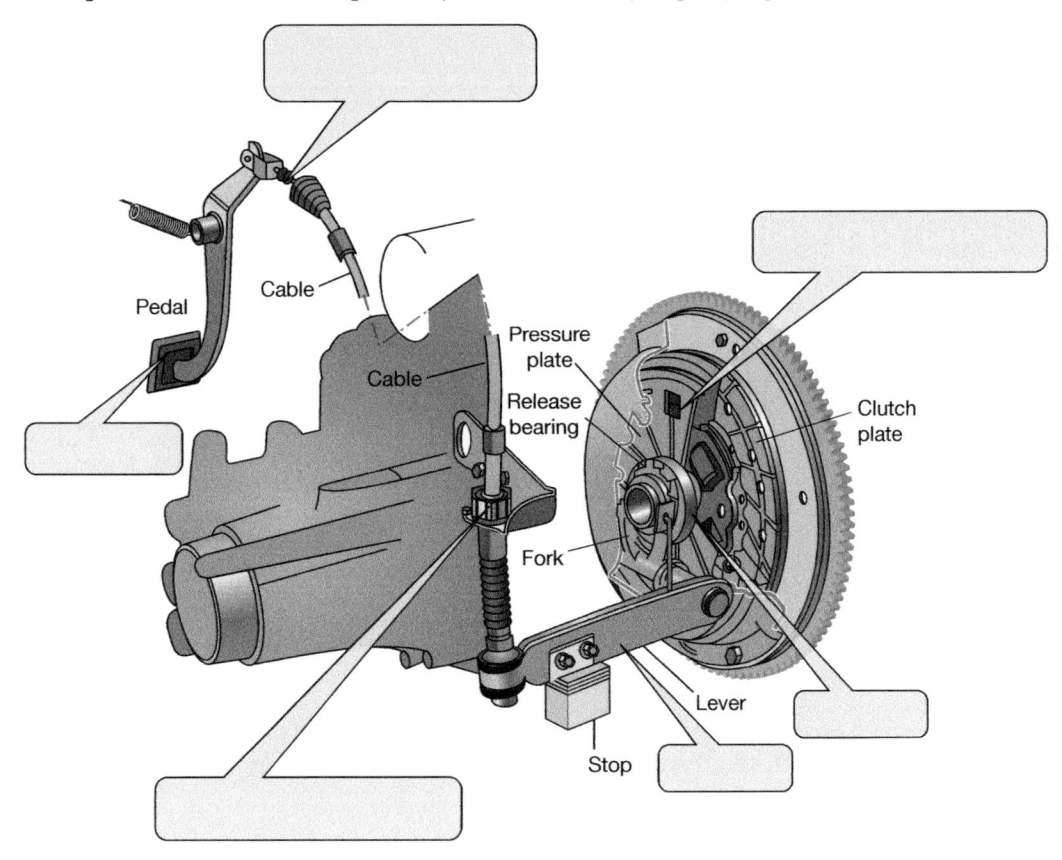

Pedal

Cable

Cable

Pressure plate

Release bearing

Clutch plate

Fork

Lever

Stop

Identify the components above.

Four of the six component labels are shown below:

clutch arm **clutch release bearing** **clutch fork** **cable free play adjuster**

Hydraulic release

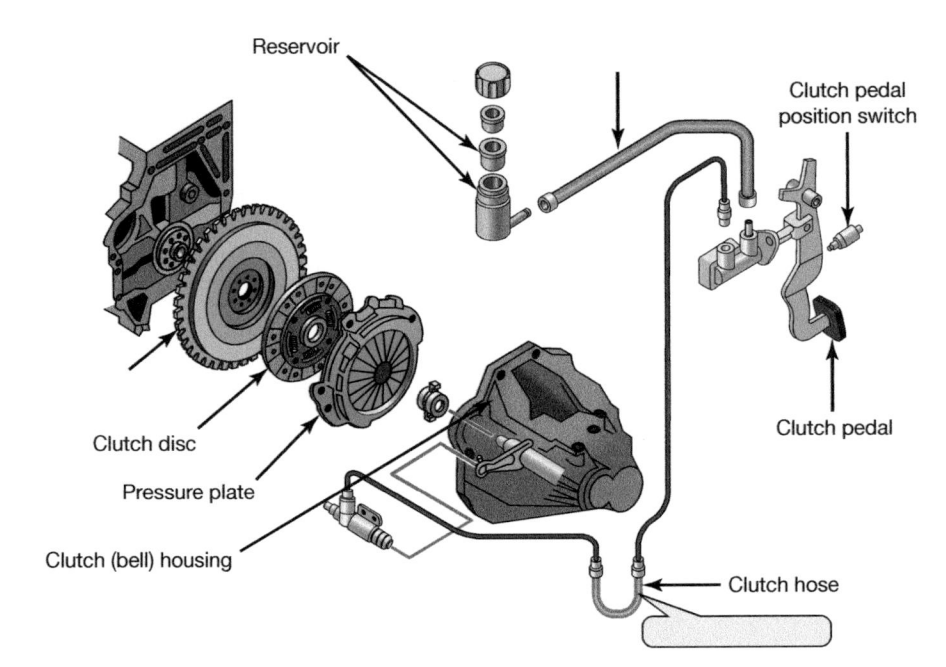

Reservoir

Clutch pedal position switch

Clutch disc

Pressure plate

Clutch (bell) housing

Clutch pedal

Clutch hose

Identify the components indicated above. Label the slave cylinder and master cylinder on the diagram above.

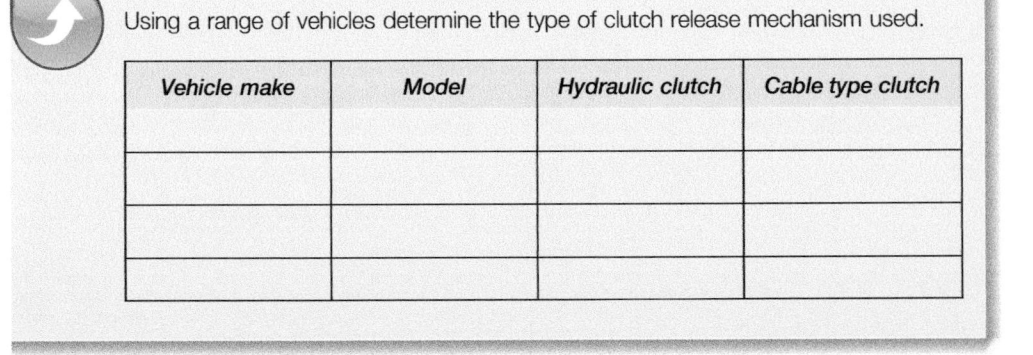

Using a range of vehicles determine the type of clutch release mechanism used.

Vehicle make	Model	Hydraulic clutch	Cable type clutch

The clutch fluid level should be checked and topped up during routine service.

GEARBOX

The gearbox is required because the internal combustion engine produces little turning force at low rpm which makes it difficult for a vehicle to pull away from rest.

Imagine trying to pull away in top gear, for example, fifth of a 5-speed gearbox.

The gearbox is able to multiply the torque to allow the vehicle to pull away easily from rest.

This is done by using a small gear driving a large gear.

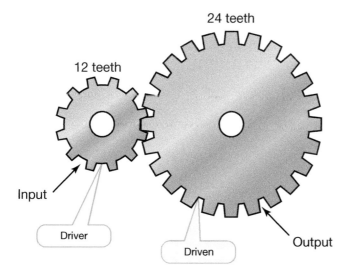

24 teeth

12 teeth

Input

Driver

Driven

Output

The input is the driving gear and the output is the driven gear.

The formula to calculate this ratio is: $\dfrac{\text{Number of teeth on the driven gear}}{\text{Number of teeth on the driver gear}}$

Calculate the gear ratio for this pair of gears:

$$\frac{24}{12} =$$

What happens to the speed of the driven gear?

_____

If the torque (turning force) going into the driving gear is 100 Nm, what will be the torque at the driven gear? _____

Using the requirements below, determine other purposes of the gearbox:

● Requirement by the driver when sitting in traffic for long periods with the engine running.

_____

● This is required when parking.

_____

There are two types of gear tooth profile used in gearboxes:

● **Spur or straight cut gears**
● **Helical gears**

Identify the gear types in this sectioned gearbox shown below.

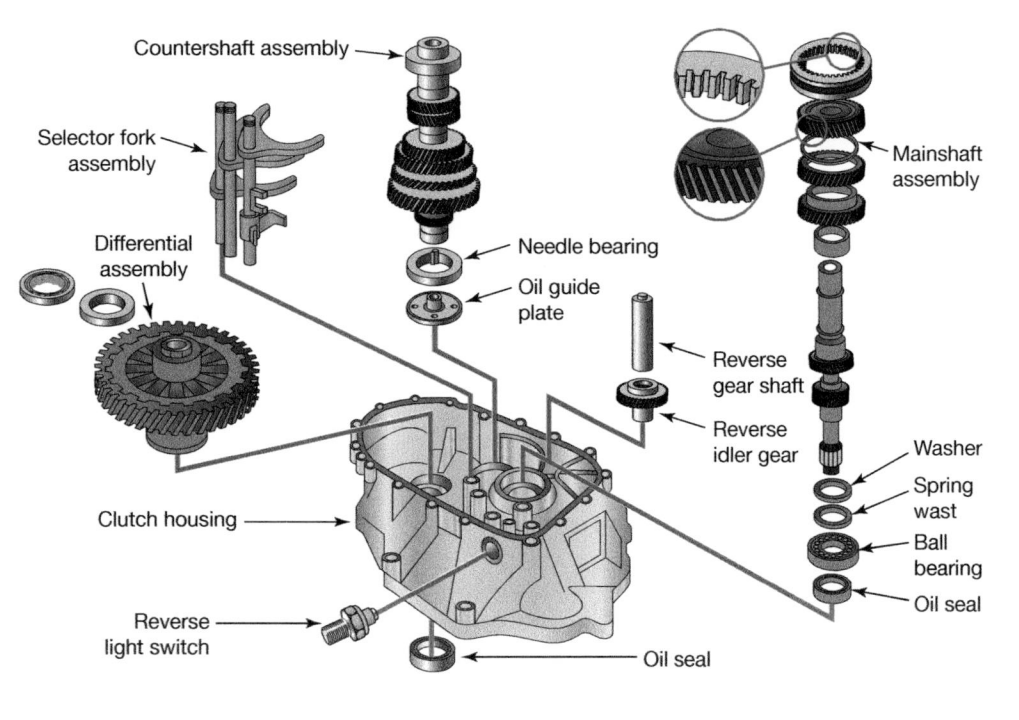

Countershaft assembly

Selector fork assembly

Differential assembly

Clutch housing

Reverse light switch

Needle bearing

Oil guide plate

Reverse gear shaft

Reverse idler gear

Mainshaft assembly

Washer

Spring wast

Ball bearing

Oil seal

Oil seal

The schematic drawing above shows the gearbox components.

In this gearbox there are two shafts containing gears.

Using the drawing above state the names of these shafts:

- _____

- _____

The diagram below shows a conventional three shaft gearbox layout. Label the reverse idler gear.

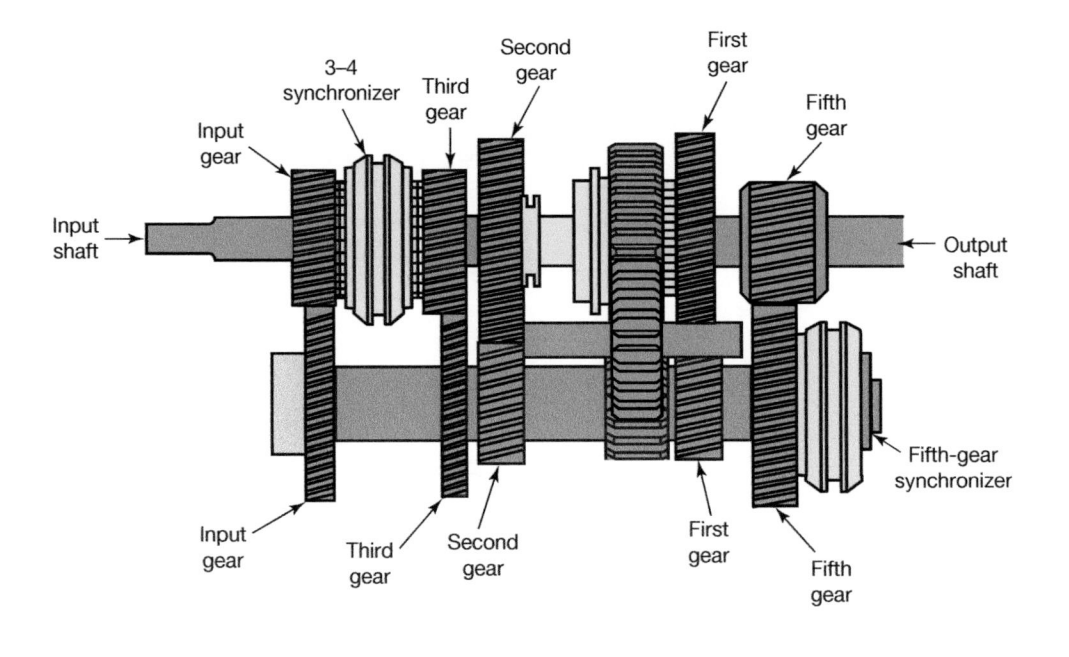

3–4 synchronizer

Third gear

Second gear

First gear

Fifth gear

Input gear

Input shaft

Output shaft

Fifth-gear synchronizer

Input gear

Third gear

Second gear

First gear

Fifth gear

Gearbox ratios

TIP A ratio is the relationship between two things, e.g. the number of gear teeth on one gear compared to another.

First gear

Shown below is the pair of gears that are used for first gear, with the input gear being the smallest with the least number of teeth. On this gearbox all forward gears have helical teeth (angled) while the reverse gears have straight cut teeth. The first gear ratio is often between 3–4:1.

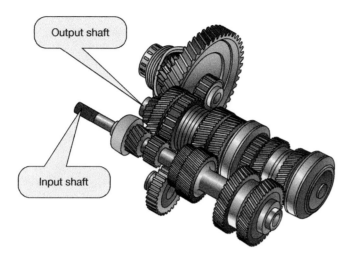

Using colours identify the following gears on the diagram above:

Second gear – blue

Third gear – green

Fourth gear – orange

Fifth gear – red

Gearbox oil level checks

RMP The gearbox oil should be checked and topped up if necessary during routine servicing.

The oil level filler plug is normally located about one-third of the way up the gearbox.

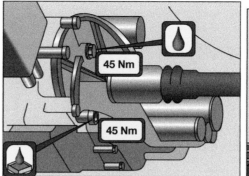
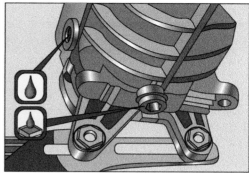

The correct grade of oil should always be used when topping up and checked using technical information.

State TWO places where this technical information can be found:

1 _____

2 _____

Procedure for checking gearbox oil with a level/filler plug

Complete the following paragraph explaining the procedure for checking gearbox oil using the words below (note: there are two extra distracter words):

level/filler	top	bottom	level
threaded hole	drain tray	tighten	drain plug

Raise the vehicle making sure it is _____. Locate the _____ plug and place a ____ _____ underneath. Remove the plug using the correct tool. The oil should be level with the _____ of the threaded hole. Top up with the correct type of oil until it is level with the bottom of the _____. Refit level/filler plug and _____ correctly.

Using the table below locate the oil level plug and the correct type of oil that is required for four vehicles with manual gearboxes.

Make	Model	Location of level filler plug	Type of oil required	Quantity of oil required

AUTOMATIC TRANSMISSION

This type of gearbox removes the need for the driver to change gears or operate a clutch. Therefore a vehicle with this type of gearbox will not have a clutch pedal. Instead of the usual friction clutch, the drive from the engine to the automatic gearbox is transmitted via oil circulation in a torque converter.

With automatic transmission the vehicle will creep forward when put in gear if the brakes are not applied.

Ensure the gear selector is in park to lock the transmission when working on the vehicle with the engine running.

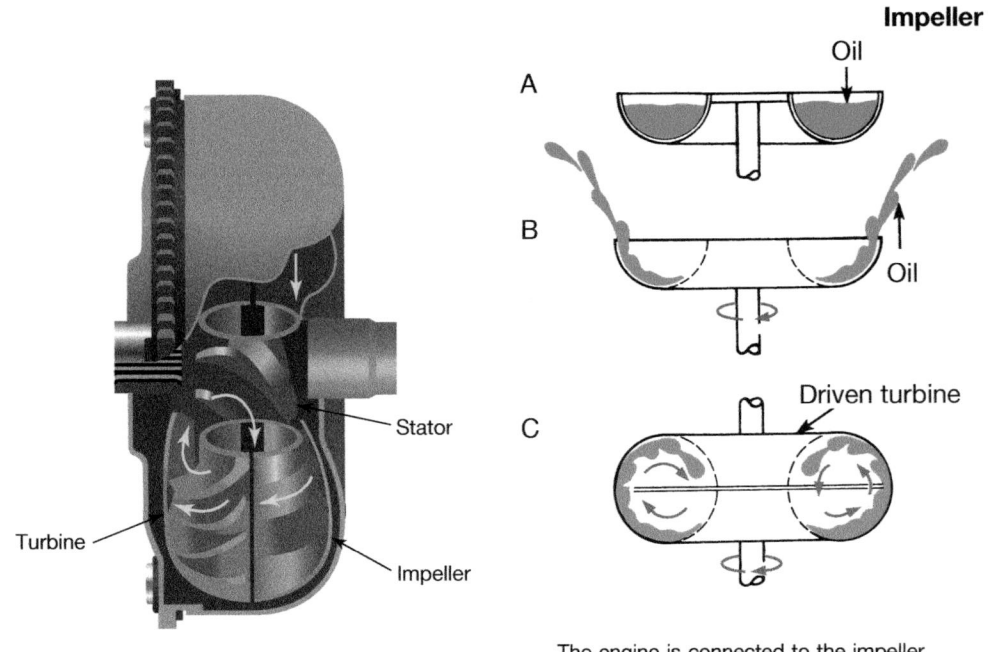

Fluid flywheel

To get an idea of how a torque converter works it is essential to know about the less complicated origins of a fluid flywheel.

On an automatic transmission the clutch is replaced by a fluid coupling relying on centrifugal force causing automatic transmission fluid to make the link between the engine and the gearbox.

Action of a fluid flywheel

The engine is connected to the impeller. The gearbox input shaft is connected to the turbine

A torque converter consists of three main parts: the impeller, the turbine and the stator. As the engine rotates, the **centrifugal** force increases causing oil in the turbine to be flung out. The faster the impeller rotates the more the centrifugal force increases. The oil is then directed into the turbine which has vanes to cause it to turn.

A similar effect can be proved using two electric fans, one running and one non-powered fan, facing each other. What do you notice?

Centrifugal force

A force that acts outward from the centre when rotation occurs, think of a playground roundabout.

Torque converter

To increase the effect of the oil moving to the turbine vanes causing torque to be transmitted more effectively, a stator is fitted in the centre to redirect the oil to improve the turning effect of the turbine.

Name the following components Impeller, Stator and Turbine.

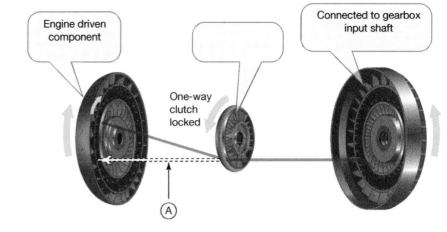

Engine driven component

Connected to gearbox input shaft

One-way clutch locked

(A)

Shade the turbine and gearbox input shaft in blue.

Shade the impeller in red.

Shade the stator in green.

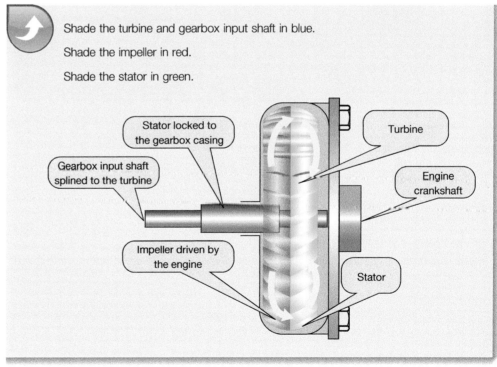

Stator locked to the gearbox casing

Turbine

Gearbox input shaft splined to the turbine

Engine crankshaft

Impeller driven by the engine

Stator

www

http://auto.howstuffworks.com

http://www.autoshop101.com

http://www.youtube.com – there are some good animations on YouTube

Complete the description of a torque converter using the word bank below:

footbrake	stator	smoothly	coupling point
turbine	centrifugal	multiplication	impeller

When the engine is idling the impeller is turning slowly and there is minimal _____ force,

therefore the amount of oil directed to the _____ is small. This allows the gearbox to be in

drive and not move, providing the driver has the _____ applied.

When the brakes are released the turbine starts to rotate and the vehicle takes off _____.

As the impeller speed increases, the oil is forced onto the turbine blades and is redirected by the

_____ back to the impeller creating a torque _____.

Most torque multiplication is achieved when the _____ and turbine are at different speeds.

As the impeller and turbine speeds equal out, the amount of torque multiplication reduces to a

point called the _____, where they both turn at the same speed.

Checking automatic gearbox oil level

Most automatic transmissions use a dipstick for oil level checks.

Using a workshop manual state the procedure for checking gearbox oil level.

● _____

● _____

● _____

● _____

Epicyclic gear train

To enable automatic transmissions to change gear smoothly, specialized gear trains are used. Different gears, that provide torque multiplication and reverse direction, can be engaged by applying brake bands.

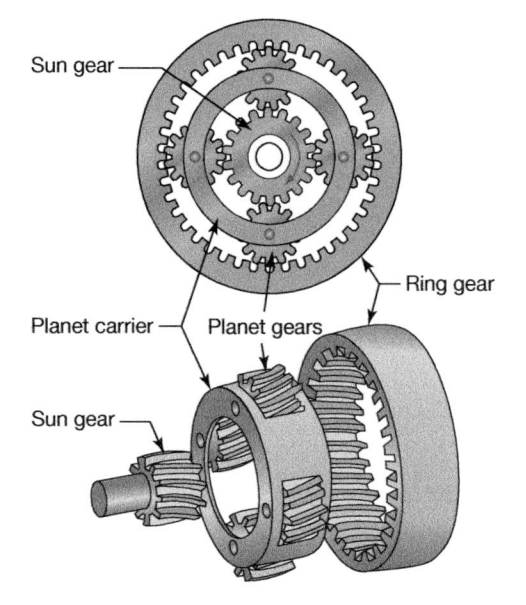

Sun gear

Ring gear

Planet carrier — Planet gears

Sun gear

Colour in the parts on the drawing using these colours:

Sun gear – Red

Planet gears - Blue

Ring gear – Black

Planet carrier – Green

What are the advantages and disadvantages of automotive transmission, compared with a manual gearbox?

Advantages: _____

_____

Disadvantages: _____

_____

FINAL DRIVE

The final drive is fitted between the drive from the gearbox to the drive to the road wheels.

It has the following purposes:

- **A final gear reduction**
- **To act as a differential**
- **To turn the drive through 90 degrees**

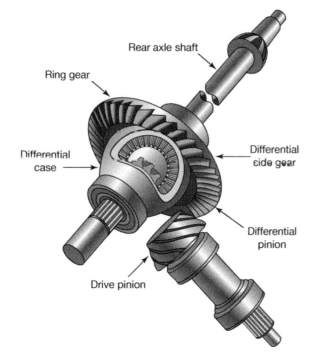

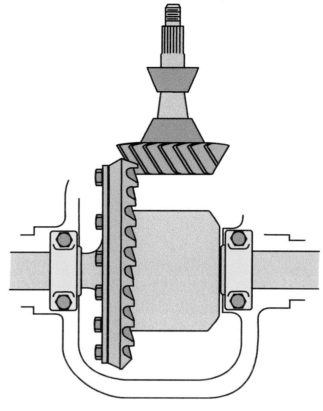

Final gear reduction

To increase torque a small gear drives a larger gear. The gear driving is called the pinion, the larger driven gear is called the crown wheel.

Label the crown wheel and pinion on the above drawing.

Typical final drive ratio is approximately 4:1.

Differential

While cornering, the outer road wheel must rotate faster than the inner wheel.

Why is this required?

_____

Complete the statement below using the following word bank:

radius   **differential**   **slow**   **smaller**   **increase**

The outer wheel needs to _____ in speed as it has to follow a larger _____, while the inner

wheel needs to _____ down as the radius is _____. A set of gears called the _____

are used to enable this speed change.

http://auto.howstuffworks.com

http://www.drivingfast.net

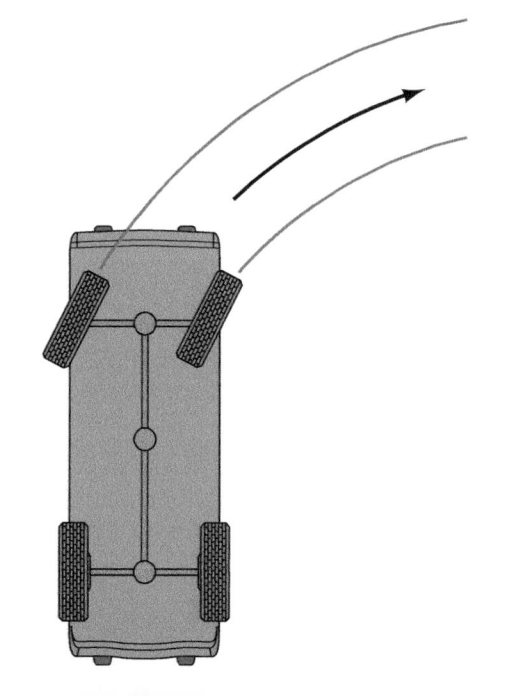

The differential is made up of:

● **Sun gears (splined to the half shafts or drive shafts)**
● **Planet gears**
● **Differential case**
● **Planet pins**

Label the diagram using these terms.

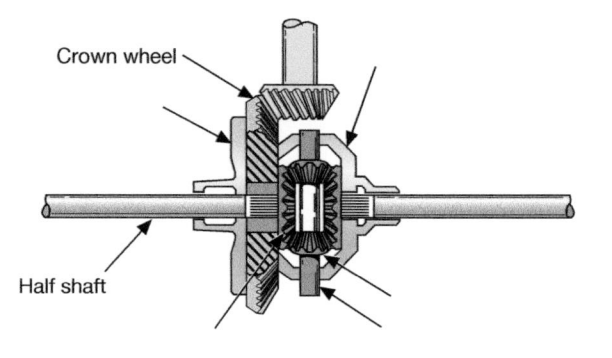

Crown wheel

Half shaft

Turning the drive through 90 degrees

On front engine rear-wheel drive vehicles the final drive is driven by the propeller shaft and turns the drive through 90 degrees to shafts connected to the road wheels.

This illustration shows the final drive assembly from a car with independent rear suspension. What is connected to the flanges labelled A, B and C?

A   _____

B   _____

C   _____

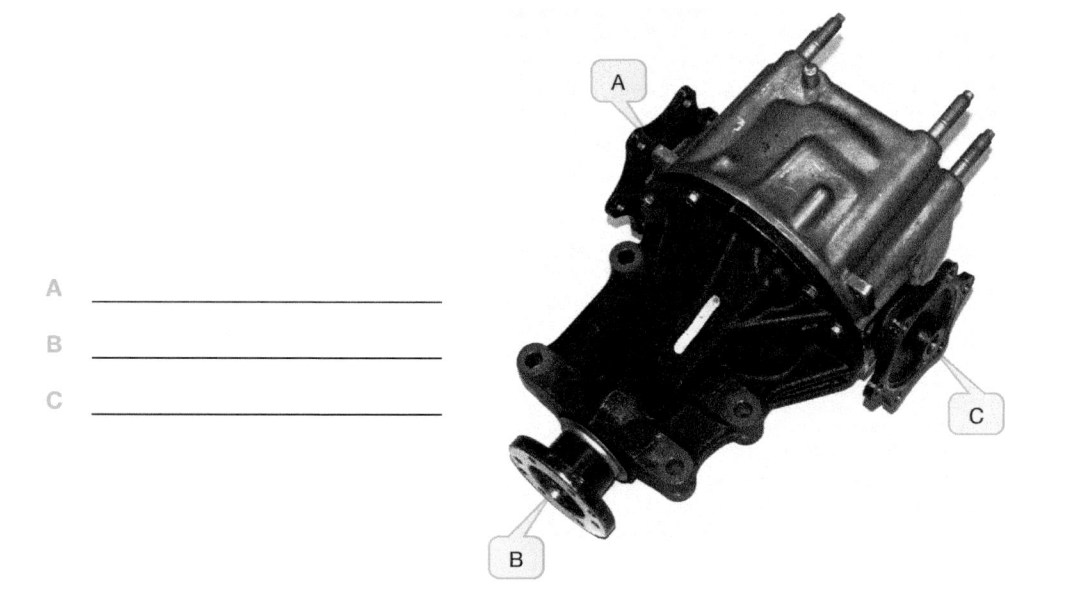

FRONT-WHEEL DRIVE

Complete the missing words. The first letter of each word is given for you.

The inboard joint is connected to the s_____ gear in the differential housed in the f_____

d_____ unit. The outboard joint has to allow for s_____ and suspension movement, while

transmitting drive to the w_____.

The inboard joint has to allow for changes in shaft l_____ as the suspension operates during all
driving conditions.

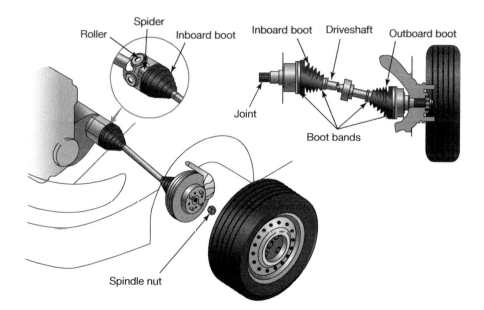

Spider
Roller
Inboard boot
Inboard boot   Driveshaft   Outboard boot
Joint
Boot bands
Spindle nut

DRIVESHAFT JOINTS

Constant velocity joint (CV joint)

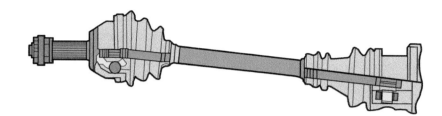

Circle the constant velocity joint in the diagram of the drive shaft above.

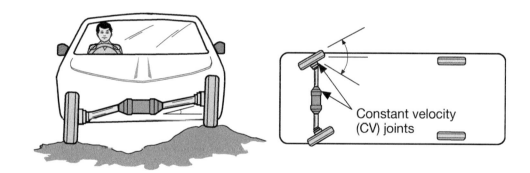

Constant velocity
(CV) joints

Describe the routine checks to be made on the driveshaft shown below.

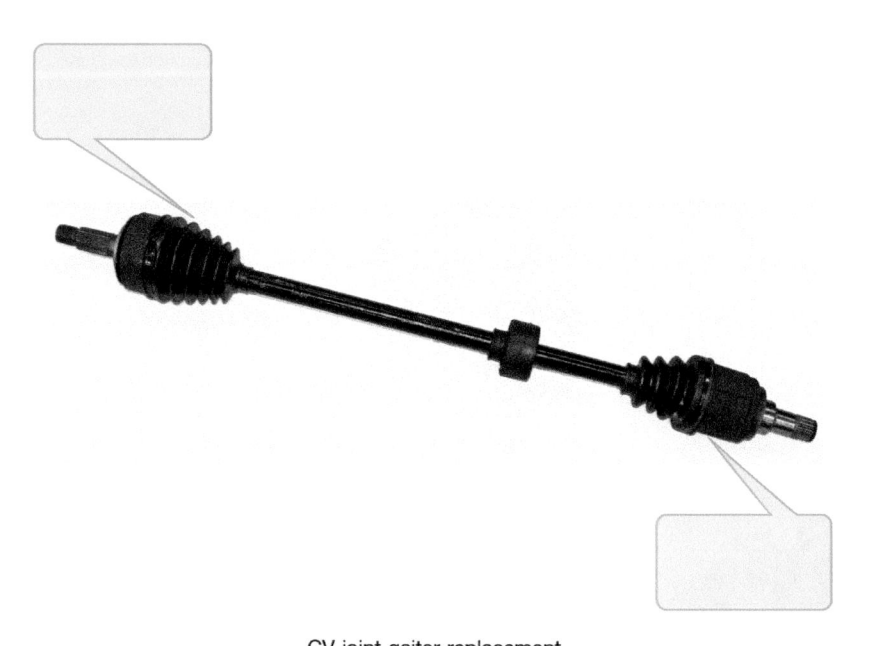

CV joint gaiter replacement

Using technical information for your chosen vehicle, describe the procedure to renew the CV gaiter.

Vehicle: Make:

_____

_____

_____

_____

_____

_____

_____

REAR-WHEEL DRIVE

Propeller shaft (shortened to 'prop-shaft').

Complete the missing words. The first letter of each word is given for you.

The prop-shaft connects the g_____ output to the final drive assembly. The prop-shaft is a hollow tube which is strong and reduces w_____; it is sometimes supported by a bearing in the centre to reduce w_____ and vibration. To allow for changes in l_____ due to suspension m_____ a sliding joint is included. **Universal joints** are used to allow for small changes in a_____ movement.

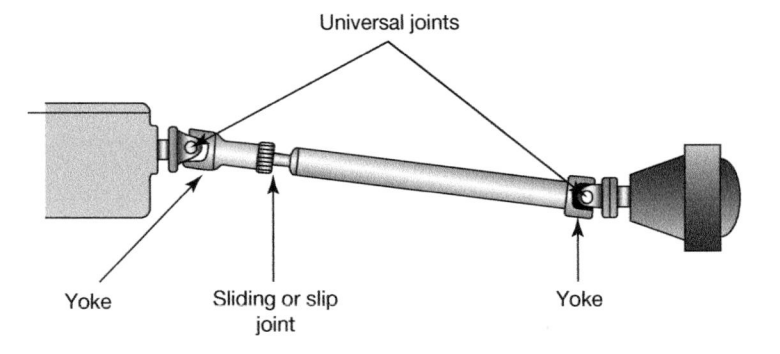

Universal joints

Yoke   Sliding or slip   Yoke
                   joint

Label the diagram of the propeller shaft above.

State THREE prop-shaft checks:

1 _____

2 _____

3 _____

FINAL DRIVE OIL LEVEL CHECK

Similar to manual gearbox check (page 185). The level/filler plug is located about a third of the way up the final drive casing.

What is this plug for?

_____

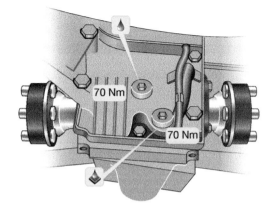

Multiple choice questions

Choose the correct answer from a), b) or c) and place a tick [✓] after your answer.

1 **Which of the following best describes a transverse engine layout?**

   a) When the engine is fitted lengthways with rear-wheel drive [ ]

   b) When the engine is fitted crosswise or sideways [ ]

   c) This term is only used when the engine is fitted at the rear [ ]

2 **What is used to transmit the drive from the gearbox to the rear axle on a front engine rear-wheel drive vehicle?**

   a) Propeller shaft [ ]

   b) Half shaft [ ]

   c) Drive shaft [ ]

3 **What component is shown?**

   a) Clutch centre plate or clutch plate [ ]

   b) Drive plate [ ]

   c) Pressure plate [ ]

4 **Which clutch component drives the gearbox input shaft?**

   a) Pressure plate [ ]

   b) Flywheel [ ]

   c) Clutch plate [ ]

5 **What is the purpose of the gearbox?**

   a) To multiply torque [ ]

   b) To allow the vehicle to pull away from rest [ ]

   c) To disengage the drive from the engine to the wheels [ ]

6 **Which component is shown here?**

   a) Torque converter [ ]

   b) Flywheel [ ]

   c) Pressure plate [ ]

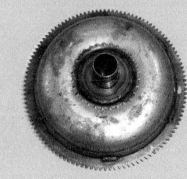

7 **Which of the following is a service requirement on a drive shaft?**

   a) Remove and inspect the CV joint [ ]

   b) Check gaiters for splits and cracks [ ]

   c) Replace gaiter clips [ ]

8 **What is a typical final drive ratio?**

   a) 2:1 [ ]

   b) 3:1 [ ]

   c) 4:1 [ ]

9 **When is a differential action used in a final drive assembly?**

   a) When travelling in a straight line [ ]

   b) When rounding a bend [ ]

   c) Going downhill [ ]

10 **What is the inboard drive shaft joint splined to?**

   a) Planet gear [ ]

   b) Gearbox output shaft [ ]

   c) Sun gear [ ]

PART 5
ELECTRICAL SYSTEMS

USE THIS SPACE FOR LEARNER NOTES

_____

_____

_____

_____

_____

_____

_____

SECTION 1
Electrical systems 195

SECTION 1

Electrical systems

USE THIS SPACE FOR LEARNER NOTES

_____

_____

_____

_____

_____

_____

Learning objectives

After studying this section you should be able to:

● Understand vehicle electrical systems and electrical principles.

● Understand how to make simple electrical circuits.

● Work safely on vehicle electrical systems.

Key terms

Alternating current (AC) Electricity that moves in two directions.
Current The flow of electricity.
Direct current (DC) Electricity which moves only in one direction.
Electrolyte A mixture of distilled water and sulphuric acid which is used in a battery.
Volt The unit of measurement for electrical pressure.
Amp The unit of measurement for electrical current flow.
Ohm The unit of measurement for electrical resistance.
Watt The unit of electrical power.

www.boschautoparts.com

www.lucaselcotrical.co.uk

www.autoshop101.com

http://phet.colorado.edu

```
E U Q R O T E O C A B L E
E L N T A T N E R R U C E
T N E C S E D N A C N I T
Y O G C H E A D L A M P L
L I O R T E T B T S N A V
O N L Y E R A S E G C O Y
R I A A B T I A R I L M N
T P H L T S R C N T H O D
C B U E E T V A A C A T I
E B R R H B H G T L C O O
L Y L Y L C E I O S I R D
E N O N E X W L R I D A E
L C U M T S T T T N E T
```

XENON   MECHANICAL   EARTH
ALTERNATOR   ELECTRICAL   SWITCH
STARTER   PINION   MOTOR
RELAY   BELT   HALOGEN
BULB   CURRENT   INCANDESCENT
BATTERY   VOLTAGE   TORQUE
ACID   RESISTANCE   HEADLAMP
ELECTROLYTE   CABLE   DIODE

ELECTRICAL PRINCIPLES

Electricity is used in all aspects of our lives. We have now come to take it for granted. Today it is used extensively in the modern motor vehicle, which means that as a vehicle technician it is important to understand how it works.

Insert the missing words into the following paragraph using the word bank below (note: there are two extra distracter words):

atoms	molecules	oxygen	negatively
positively	nucleus	two	hardly
three	proton	centre	hydrogen

Everything in our world is made up of _____ and the molecules in a substance are made up from _____. Water is made up from molecules and atoms. A molecule of water, H_2O, contains _____ atoms of _____ and one atom of _____.

If we look at one hydrogen atom there is a _____ (proton) in the _____, which is _____ charged, and one electron, which is _____ charged, circling around the _____.

The circle below shows the orbit of an electron in a hydrogen atom. Complete the diagram showing the electron and proton, stating which is positively and which is negatively charged:

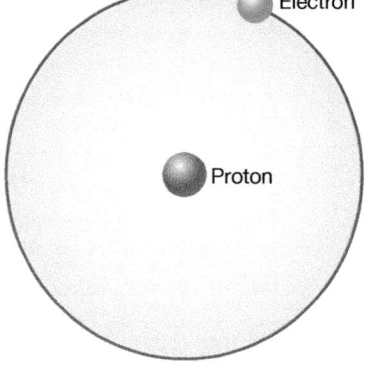

A hydrogen atom

To make an electric current flow the electrons need to be moved from one atom to the next. The current needs to have a force applied to it. Name one component in a vehicle which applies this force: _____

Copper, which is a very good conductor, contains 29 protons and 29 electrons.

A conductor offers very little resistance to the flow of electrons. Good conductors are mainly metals, such as c_____, a_____, g_____, i_____ and s_____.

A non-metallic material which is also a very good conductor is _____.

Insulators are materials that resist the flow of electrons and are ideal for insulating wires and cables. Good insulators include p_____, r_____, g_____, p_____ and c_____.

Magnets have a positive and negative pole which are known as: _____

Like poles _____

Unlike poles _____

Motors use this principle to make them work. This is because the magnetic effect can be used to make things move.

If an electric current is passed through a conductor like a copper wire, it will generate a magnetic field. This magnetic field is invisible.

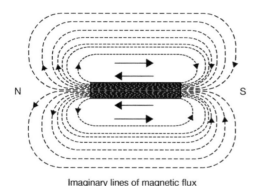

Imaginary lines of magnetic flux

If a length of copper wire is passed by a magnet then the magnetic attraction will cause an electric current to flow in the wire.

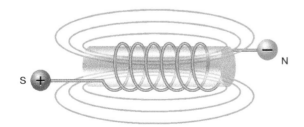

It can be difficult to understand how electricity works, as it is something we cannot see. A simple and easy way to understand electricity is to look at water and how it flows in pipes and hoses.

Tank 'A' is a reservoir –
  representing the battery (fully
  charged, positive)
Tank 'B' is another reservoir –
  representing the battery
  (negative)
The hose – represents the
  conducting wire
The tap – represents a switch
The water wheel – represents a
  motor

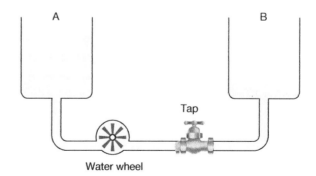

The diagram above shows two water tanks connected by a water hose. Tank 'A' is full of water. There is a tap which can be opened and closed to control the flow of water and a water wheel which rotates when water is flowing. Because the tap is closed there is no flow of water.

In the diagram above the tap is half open, water begins to flow due to the pressure of the water in tank 'A' and the water wheel rotates. The more the tap is opened the quicker the water will flow and the faster the water wheel will rotate.

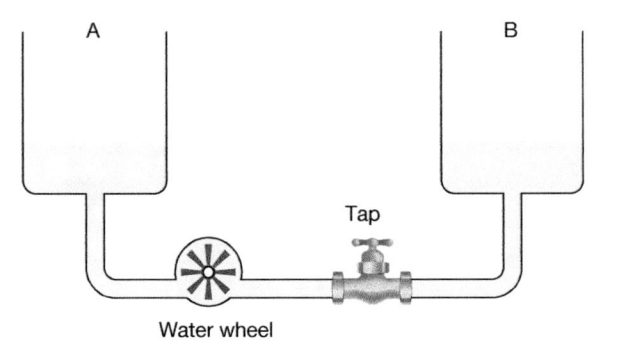

A   B

Tap

Water wheel

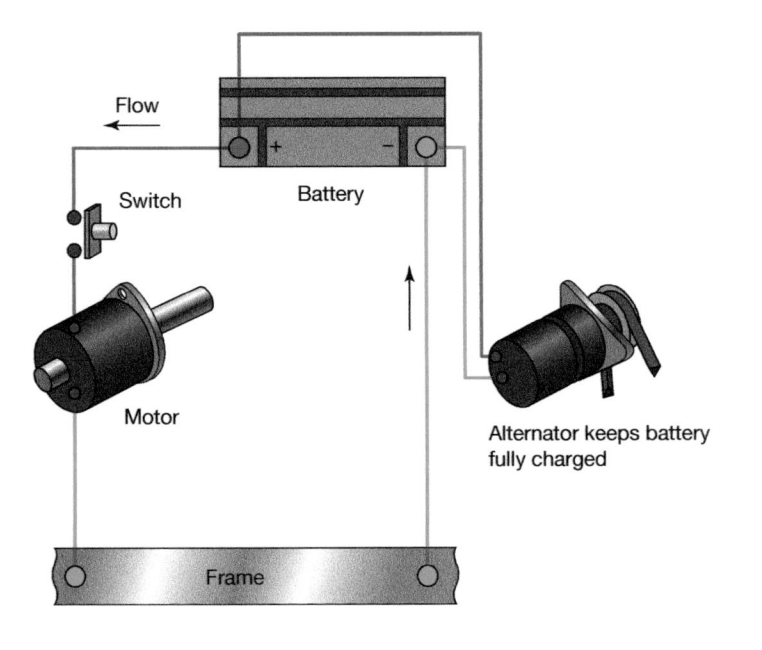

Flow

Switch   Battery

Motor

Alternator keeps battery
fully charged

Frame

Now when the tap has been left open for a long while, as in the diagram above, the pressure in tank 'A' becomes less until the water stops flowing and the wheel stops rotating.

Lets us now compare the water tank circuit to electricity and an electrical circuit. Draw a line from the water tank circuit comparison to the correct electrical comparison. Use the diagram opposite to help you:

Water tank 'A'		This is the resistance in the circuit, which can be varied to control the flow of current

Water wheel		Current

Tap		A motor which is powered by the flow of the water

The flow of water		Voltage–the pressure which pushes the electricity through the circuit. This could be the battery

ELECTRICAL UNITS OF MEASUREMENT

There are four main electrical units of measurement.

State the electrical units and symbols for the following:

Pressure: _____

The pressure is in the circuit when it is switched on and the current can flow through the circuit.

When the pressure is stored it is known as EMF,or E_____ M_____ F_____.

Current: _____

This is the amount of electricity flowing in a circuit at any given point.

Resistance: _____

When something in an electrical circuit restricts the flow of electricity, such as a bulb, it creates an opposition to the current.

Power: _____

This is the energy given up by the flowing electricity in a circuit. For example, when a light is switched on it will glow brightly and generate heat.

Ohms Law

There is a relationship between voltage, current and resistance. If the voltage (pressure) is increased in a circuit, more electricity would flow and the amperage (current flow) would also increase.

If the resistance in the circuit is increased, then the amperage (current flow) would decrease. This relationship is known as Ohms Law.

The symbols used are:

V = _____

R = _____

I = _____

The first two are self-explanatory, but the last symbol can be a bit confusing. It is thought that 'I' is used because it represents the 'Intensity' of current flow. Sometimes you may see 'E' being used to show voltage, which stands for 'electromotive force'.

A triangle is used to show the relationship between V, I and R. We need to have two of these values in the triangle to be able to calculate the missing value.

Complete the following triangle by inserting these symbols (V, I, R):

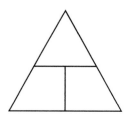

With the completed triangle, cover up the missing or unknown circuit value. The other two known values are either multiplied or divided to find the unknown value. Complete the following mathematical calculations:

_____ = amps × _____

amps = _____ × _____

_____ = volts ÷ _____

The following example shows a circuit with a current of 10 A and a resistance of 1.2 Ω, where the missing value to calculate is voltage:

$$V = I \times R$$
$$V = 10\,A \times 1.2\,\Omega$$
$$V = 12\,V$$

Now calculate the circuit resistance when the pressure is 12 V and the current flow is 4 A:

_____

_____

_____

Power triangle

By using a similar method to Ohms Law, watts and power can be calculated.

The symbols used are:

P = power in watts (sometimes shown as 'W')
V – volts (sometimes shown as 'F')
I = amps

Complete the following power triangle:

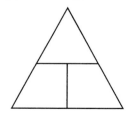

With the completed triangle, cover up the missing or unknown circuit value. The other two known values are either multiplied or divided to find the unknown value. Complete the following mathematical calculations:

$$\text{watts} = \underline{\hspace{2cm}} \times \underline{\hspace{2cm}}$$

$$\text{volts} = \underline{\hspace{2cm}} \div \underline{\hspace{2cm}}$$

$$\underline{\hspace{2cm}} = \text{watts} \div \text{volts}$$

Circuit symbols

When working with electrical systems it is common practice to use circuit diagrams. These are like electrical maps which show the various components, cable sizes, component locations and pin terminal numbers. They are useful to use when fault-finding.

In the table below, name the common electrical symbols used in circuit diagrams:

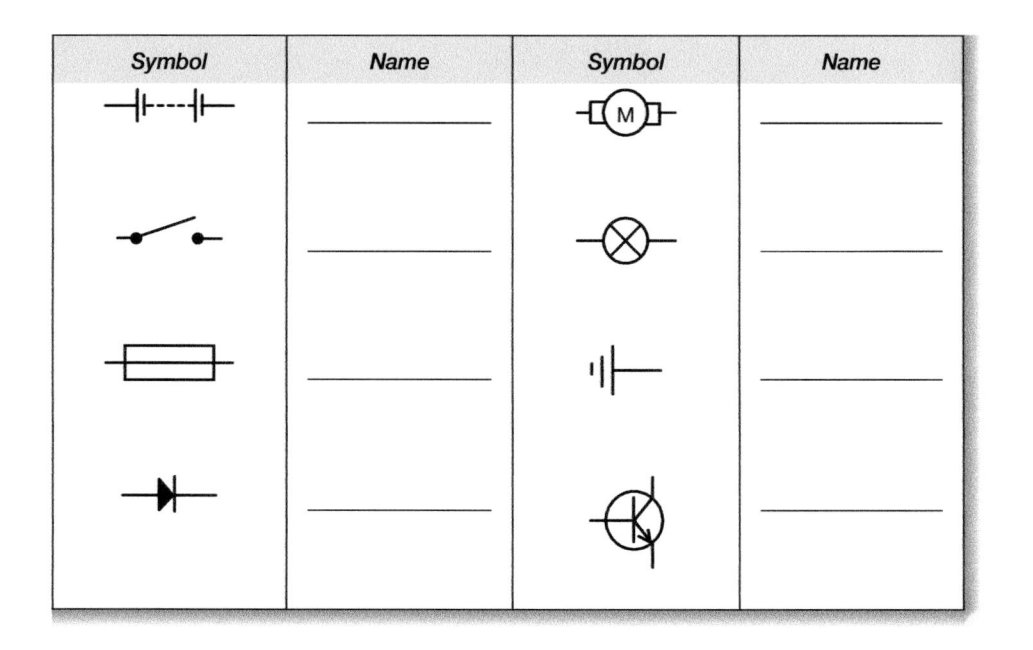

Symbol	Name	Symbol	Name
⊣⊢---⊢	_____	⊢(M)⊢	_____
•—•/—•	_____	—⊗—	_____
▭	_____	⊣⊢	_____
—▷⊢	_____	⊕	_____

A relay is a common component used in motor vehicles. Sketch the symbol for a relay:

Research and find the Japanese electrical symbols used in circuit diagrams. They differ from those shown in the table.

SIMPLE ELECTRICAL CIRCUITS

For an electrical circuit to be complete it must have an unbroken loop so that current can flow, from the source, around the circuit, and back to the source. For example, from a battery via a cable to a bulb and back again to the battery.

Give examples for each of the following circuit components:

A power source: _____

Connections: _____

Consumers: _____

On/off control: _____

Circuit protection: _____

Circuit types

There are two main types of electrical circuits, both of which are used in motor vehicles. These are series and parallel.

Complete the following diagram for each of the circuit types:

Series circuit   Parallel

Earth return and insulated systems

Insert the missing words into the following paragraph using the word bank below (note: there are two extra distracter words):

earth	positive	reduce	circuit
power	chassis	wiring	metal
two	three	returns	connecting
negative	complete	battery	

For an electrical _____ to be complete it must have some form of _____ return back to

the _____ source (battery). This is normally achieved by one of _____ methods.

Most modern vehicles have their _____ made from _____. To help _____ the

amount of _____ a majority of vehicles use the chassis to _____ the electrical circuit.

The _____ (earth) of the vehicle _____ is connected to the chassis. A vehicle circuit is

completed by _____ its earth to the vehicle chassis, so that it _____ back to the

negative of the battery.

Complete the following diagram of an earth return system. Correctly label the components:

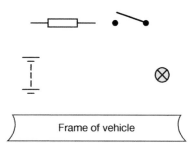

Frame of vehicle

Some vehicles use insulated earth return. How does this differ from the chassis earth return?

_____

_____

Name a type of vehicle which uses insulated earth return and state why this is considered essential:

_____

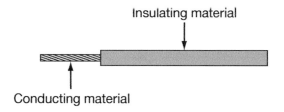

Insulating material

Conducting material

Basic construction of a cable

Vehicles use a lot of cables for their many circuits, such as the brake light circuit and the horn circuit. To make it easier to identify the different circuits the cables are colour coded via their insulators. It is normal for the wire colour code to be shown on wiring diagrams. Some manufacturers may use numbers as well as, or instead of, colours to aid identification.

The cables need to be protected from touching each other. This is achieved by insulating them, and it is this insulation that is usually colour coded.

Insert the missing words into the following paragraph using the word bank below (note: there are two extra distracter words):

conducting	thickness	conductor	cables
number	thicker	insulator	overheat
strands	more	copper	circuit
electricity	number	failure	
resistance	size	fire	

Cables are rated by the _____ and _____ of the wire strands that they have. The figure on page 201 shows the basic construction of a cable. The _____ material is usually made up from a number of _____. The _____ strands and the _____ the cable, the more _____ it can carry. _____ is the common _____ used in these wires and _____. It is important to use the correct _____ cable for the electrical _____ which it is intended for. If a cable does not have the _____ of strands at the correct thickness, it could _____ and melt, causing electrical _____ and possibly a _____.

It is important that the correct wire is chosen and used for an electrical circuit. The table below gives some examples of wire sizes and their uses in a motor vehicle.

Number of strands/wire diameter (mm)	Typical usage and continuous current rating
9/0.30	Horn, side lamps, tail lights, reversing lights 5.75 A
65/0.30	Charging cable for an alternator 35 A
37/0.71	A starter/battery cable 170 A

When electrical current flows through cables, heat is generated. This heat can be used in many ways. Give THREE examples of where this is effectively used in a motor vehicle and briefly explain their function:

1 _____

2 _____

3 _____

RMP Visually and physically check the security and condition of all exposed cables and wires.

CIRCUIT FAULTS

During the normal working life of a vehicle, faults in the electrical circuits are likely to occur. Modern vehicles are generally very reliable.

Two circuit faults are shown below. Correctly label each from the following:

Short circuit, Open circuit

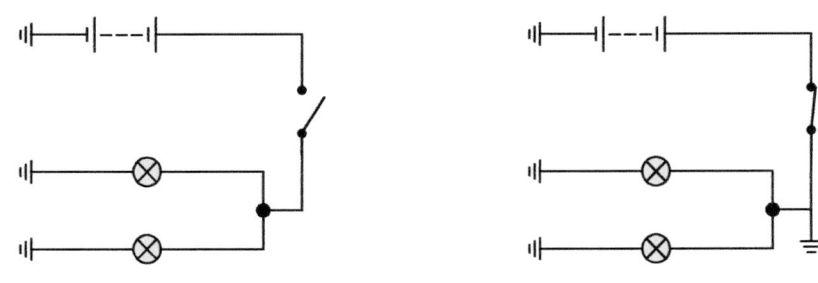

_____   _____

What should be fitted to a vehicle electrical circuit that is designed to help protect it from a short circuit? _____

Another type of electrical circuit fault is when the current meets unexpected restriction in the circuit which reduces its ability to flow. This is known as a _____

On the following table draw lines from the fault to the correct symptom:

Fault		Symptom
Open circuit		When the headlights are switched on they are dim
Short circuit		When a switch is operated nothing works
High resistance		The fuse keeps blowing

MULTIMETER

Electrical circuits may be checked using a piece of test equipment known as a multimeter.

The two types of multimeter used are:

_____: These show the electrical readout by means of a

moving needle on a screen. They can be difficult to read accurately.

_____: These show the electrical readout in the form

of numbers on a liquid crystal display (LCD). These are very accurate

and are the most commonly used type of multimeter today.

There are two types of digital multimeter. These are:

_____: These are normally operated by manually

turning a dial to select the correct unit (volts, amps, ohms) and scale to be used.

_____: The correct unit is manually selected (volts, amp, ohms) and the scale is

selected automatically by the multimeter.

There are two types of current measured in volts and amps. These are DC (_____)
and AC (_____). Draw the correct electrical symbol for each of these below:

AC   DC

Picture supplied by Draper Tools Ltd

Multimeter, digital,
automotive, auto-ranging

AMMETER

This is used to measure the electrical _____ in a circuit. It can be used as part of a multimeter function by turning the selector dial to measure amps.

OHMMETER

This is used to measure the electrical _____ in a circuit. It can be used as part of a multimeter function by turning the selector dial to measure ohms.

Using a multimeter

The leads need to be connected correctly to the circuit. The three diagrams below show a DMM (_____) correctly connected to check resistance, current and voltage. Label the diagrams with the check being carried out.

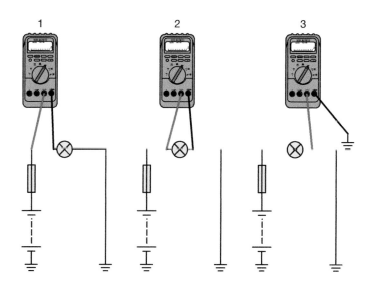

When using the DMM for measuring electrical circuits and components, the leads need to be connected correctly to the meter. Complete the following table:

DMM connection and settings	Reading
Black lead connected to 'Comm' socket, red lead connected to 'V/Ω' socket. Dial set to read:	_____
Black lead connected to 'Comm', red lead connected to 'A'. Dial set to read:	_____
Black lead connected to '_____' socket, red lead connected to '_____' socket. Dial set to read:	Resistance.

BATTERY

Label parts a–d on the diagram of a typical 12 V battery below.

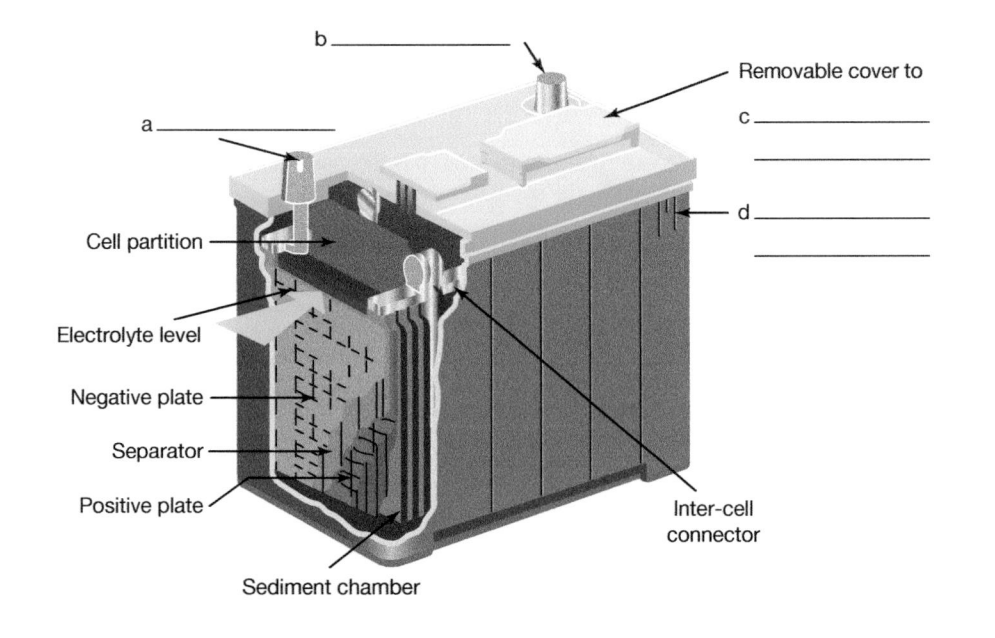

b _____

a _____

Cell partition

Electrolyte level

Negative plate

Separator

Positive plate

Removable cover to

c _____
_____

d _____
_____

Inter-cell connector

Sediment chamber

Vehicles need to have a means of storing electrical energy to supply power for components such as the starter and lights when the electricity is not being generated by the engine. This is achieved by using a battery.

The most common type of battery used in a motor vehicle is a _____

Electrical power is important for a vehicle to work. The modern vehicle has a lot of electrical consumers. List FIVE such systems:

1 _____

2 _____

3 _____

4 _____

5 _____

All of these need a reliable source of electricity to power them. This is where modern batteries are designed so that they can be recharged. The component that recharges them is known as the

_____.

The active plates in this type of battery are made from two types of lead which are immersed in a mixture of sulphuric acid and distilled water.

What name is given to the liquid mixture in the battery? _____

Maintenance-free battery

Some batteries are maintenance-free and completely sealed. To check the state of charge, look down the indicator window on top of the battery. Inside is a built-in hydrometer which measures the specific gravity of the electrolyte. The higher the number the higher the state of charge.

Both lead acid and maintenance-free batteries contain two electrodes or plates surrounded by the electrolyte. These three elements make up an electrochemical battery cell.

Circle the correct statement below:

A 12 volt lead acid battery has 12 cells

A 12 volt lead acid battery has 6 cells

A 12 volt lead acid battery has 4 cells

Top up lead acid batteries with distilled water only.

With the use of textbooks and the Internet, research the chemical effect of charging and discharging of a lead acid battery.

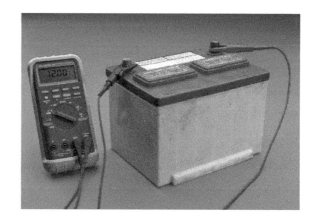

The image above shows a digital multimeter connected to a battery. What is this measuring?

_____

Sometimes batteries need to be removed from a vehicle to be recharged.

The first task is to correctly disconnect the battery. Which terminal is disconnected first? _____

Give ONE reason why this terminal is disconnected first:

_____

Now the battery is disconnected and unclamped from the vehicle. Complete the rest of the procedure:

1 Remove the battery from the vehicle.
2 Carry the battery to the bench charger.
3 Check the level of _____
4 Charge at a low charging rate.
5 When the battery is fully charged, switch off the _____. If this is not done correctly a spark could cause the battery to explode.
6 Remove the _____ from the battery.
7 Replace the charged battery in the vehicle. Clean and secure the terminals.
8 Start the engine.

Check the security, of the battery and terminals. Check for any damage to the battery casing. Check the fluid level and state of charge where possible.

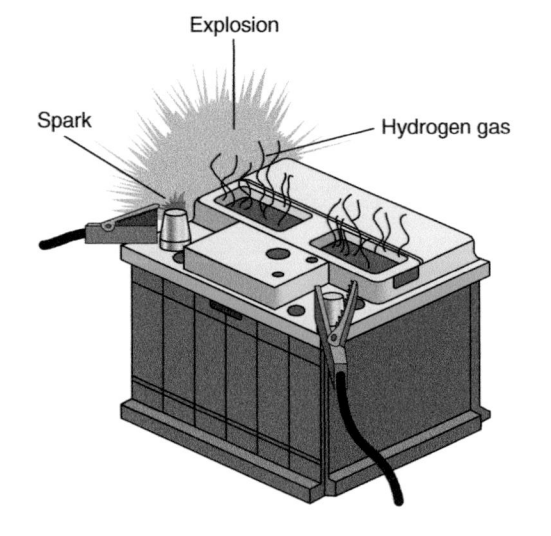

Spark

Explosion

Hydrogen gas

When carrying a battery wear gloves and overalls.

The acid can burn if spilt. Batteries are heavy, and hard to carry. The battery charging area must be well ventilated.

JOINING TWO CABLES

When working on vehicles it is always important to select and use the correct tools.

State ONE method of joining cables:

_____

These come in different sizes and are colour coded accordingly.

When selecting this type of connector you need to select the correct size for the cables in the circuit. The crimping pliers have coloured markings on them which correspond to the same coloured crimp connector.

CHARGING

Bosch

Alternator

Insert the missing words into the following paragraphs using the word bank below (note: there are two extra distracter words):

mechanical	energy	start	provides
supply	starting	parts	running
crank	charge	main	converts
drives	warning	number	belt
components	regulator	wiring	engine
charging	battery	key	

The _____ purpose of the alternator is to provide electrical _____ for all of the

vehicle's electrical _____. The _____ system consists of a _____ of main

_____. These include:

● Battery
● Alternator
● Voltage _____
● Charge _____ lamp
● Electrical _____ for the circuit.

When the ignition _____ is turned to _____ the vehicle, electrical _____ which is

stored in the _____ then _____ the electrical energy needed for _____. When the engine is _____, the alternator _____ some of the _____ energy from the _____ into electrical energy. The alternator is driven by a _____ from the _____ pulley.

The alternator also charges the battery. It is important that the alternator does not overcharge the battery or exceed a pre-set maximum voltage for charging. The component which prevents over-charging is the _____.

ALTERNATOR CONSTRUCTION

The main components of an alternator are the rotor, stator, regulator and rectifier.

Correctly label the figures that follow with these component names:

stator   **regulator**   **rotor**

Correctly label the diagram opposite.

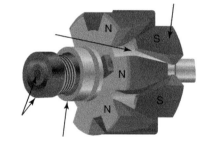

  Check the security, tension and condition of the alternator drive belt. The belt should be free from any cracks and splits.

Stator neutral junction

To diodes   To diodes

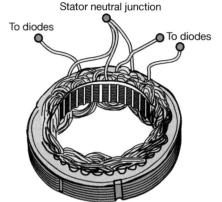

_____   _____

The rectifier is another major component used in an alternator. What is its purpose?

_____

Correctly join the component name with its function:

Rotor	Electric current is generated in these series of three coils of copper wire
Stator	Used to convert AC to DC by means of diodes
Rectifier	Controls the maximum output of the alternator
Regulator	An electromagnet that is turned by means of a pulley and belt via the engine

For an alternator to be useful in a motor vehicle it needs to produce a certain type of current. Circle the correct current that is produced:

| AC (alternating current) | DC (direct current) |

Alternators need very little maintenance.

What is the most common fault with an alternator? _____

The common symptoms of this fault are:

1 _____

2 _____

Before testing an alternator, check:

● The _____ condition and its state of _____.
● The connections between the battery and the _____.
● For breaks in the leads.
● For a loose or worn drive _____.
● Excessive signs of wear and noise.

Circle the correct statement in the table below:

An alternator converts electrical energy into mechanical energy	An alternator converts mechanical energy into electrical energy

STARTING

Bosch

Starter motor

For the engine to start, the crankshaft needs to rotate at quite a fast speed. This is achieved by the means of a heavy duty electrical motor known as a starter motor.

What is the minimum speed that the starter motor needs to turn? _____

Name TWO types of starter motor: _____

The image opposite shows a typical modern starter motor. What is the name of the silver cylindrical component indicated by the arrow? _____

A starter motor converts mechanical energy into electrical energy	A starter motor converts electrical energy into mechanical energy

List FIVE main components of a vehicle starting system:

1 _____

2 _____

3 _____

4 _____

5 _____

Insert the missing words into the following paragraphs using the word bank below (note: there are two extra distracter words):

block	operate	energized	gear
electric	cranking	high	starts
start	motor	pinion	low
ring	solenoid	ignition	stops

The starter motor is an _____ motor which is mounted on the engine _____. When the _____ key is turned to the _____ position the motor is _____ by the battery. The _____, known as the starter _____, engages with the flywheel _____ gear; the starter motor begins to _____, turning over the engine until the engine _____. This is also known as '_____' the engine.

The starter _____ is designed to have a _____ turning effort at _____ speeds.

A common name given to turning effort is _____.

Why are the cables of the starter motor the thickest on the vehicle?

_____

_____

What would happen if the cables used in the starter motor circuit were not thick enough?

_____

Complete the basic electrical diagram for the starter circuit below. Label the: battery, solenoid, main contacts, starter motor, ignition switch.

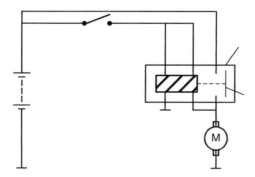

Match the components to their correct descriptions.

| Armature | This is where the magnetic field is created and is used to turn the armature |

| Field coils | The rotating central shaft in the starter motor that is turned by the magnetic field |

| Solenoid | This is the small gear which is attached to the end of the armature |

| Pinion gear | A heavy duty switch. This is used to move the pinion gear to engagement |

TIP If a single light has failed it is likely that a bulb has blown. If both headlamps, braking or parking lights have failed the problem could be that the switch is faulty or the fuse has blown.

LIGHTING

Examine a vehicle and check that the lights operate properly.

Vehicle: Make .... Model ....

Lights	Working	Not working	Action needed	Wattage
Front sidelights				
Brake lights				
Rear sidelights				
No. plate light				
Direction indicators				
Rear fog light(s)				
Headlamp main beam				
Headlamp dip beam				
Reversing light(s)				

The motor vehicle uses a number of different types of lights for various functions. Most of these lights make use of the heating effect of electricity as it passes through a wire of high resistance. The heat is evident as a form of light.

Complete the table below with the correct lens colour for each lamp:

Lamp	Colour
Headlamps	
Side lamps	
Indicator	
Brake lights	
Rear lamps	
Reverse lamps	
Front fog lamps	
Rear fog lamps	
Rear number plate lamps	

 Some, but not all, lighting needs to meet the legal requirements such as the MOT test.

RMP Check for the correct operation of warning lamps. Always check for the correct specifications in the MOT tester manual.

Lights are also used for driver information. For example, the oil pressure warning light. Name the SIX other driver information warning lights shown below that can be found in a modern motor vehicle:

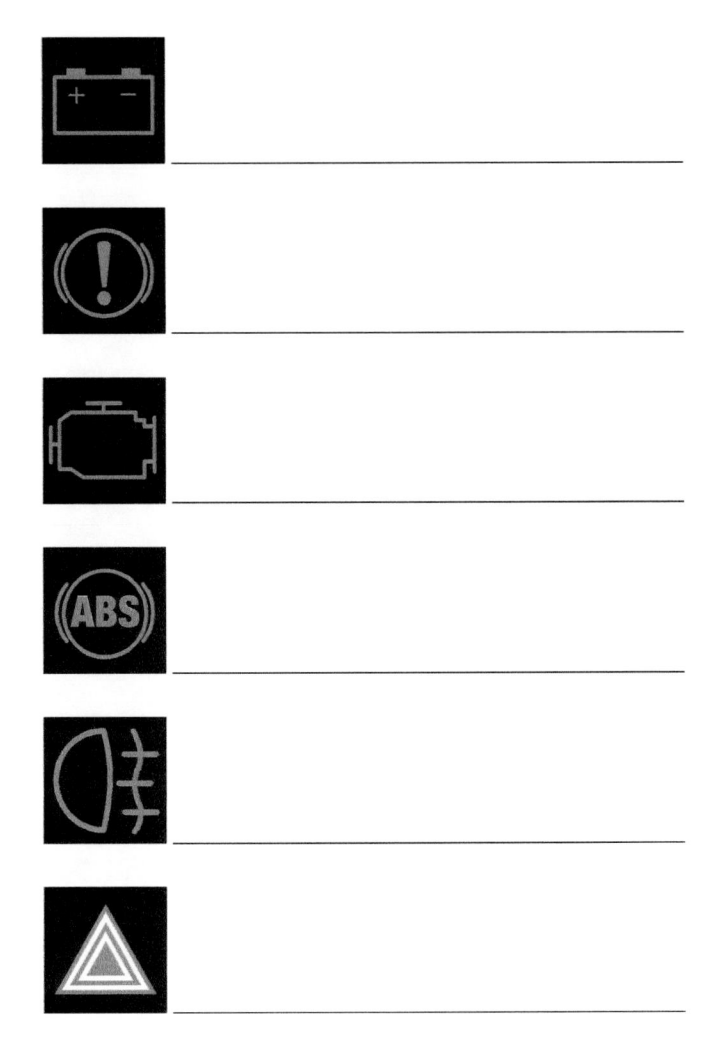

A typical instrument panel showing driver information warning lights

Insert the missing words into the following paragraphs using the word bank below (note: there are two extra distracter words):

tungsten	**bulbs**	**fog**	**stop**
off	**headlights**	**halogen**	**main**
on	**cool**	**reduce**	
quartz	**hot**		

The main types of _____ used on modern vehicles include conventional (vacuum, inert) lights.

_____ incandescent light bulbs have been used as the _____ light source in various

motor vehicle systems, including _____ lamps, _____ lamps and signalling lamps.

_____ is very commonly used in _____. When replacing a _____ halogen headlamp

bulb it must be switched _____ and _____. Avoid hand contact with the quartz envelope

(the clear casing) as this can _____ the life of the lamp.

RMP Visually check the operation of all legal lighting systems. Ensure that they are secure and of the correct colour. Check headlamp aims and patterns are correct.

Incandescent light bulbs (tungsten filament)

The tungsten incandescent light bulb has traditionally been the light source used in all of the motor vehicle lighting, signalling, marking and warning lamps.

Sketch a basic single element bulb and label with the following:

glass envelope   **filament**   **contact**   **bayonet pins**

When the tungsten filament is heated by the electric current passing through it, the filament will glow white hot. The filament is coiled into a spiral allowing a long length of wire into a small space. The glass envelope contains an 'inert' gas, such as argon. This allows the element to operate at a very high temperature, producing a stronger white light without burning out.

Look at a selection of vehicles in the workshop. Find the wattage of the bulbs fitted for the:

- Headlamps
- Side lamps
- Indicator
- Reverse lamps
- Rear lamps
- Brake lights
- Front fog lamps
- Rear fog lamps
- Number plate lamps

List the correct and safe method for replacing a light bulb:

1 _____

2 _____

3 _____

4 _____

5 _____

6 _____

Quartz halogen bulbs

These bulbs get very hot. The envelope is made from quartz (a rock crystal) instead of glass, because it can withstand higher temperatures. A pressurized gas called halogen is contained in the bulb.

What is an advantage of using a quartz halogen bulb?

_____

HID (H_____ I_____ D_____)

This type of headlamp is becoming common on modern vehicles. They are very different from the filament type bulb as they do not use an electrical filament which gets hot and glows. Instead they use two electrodes and a high voltage spark jumps between them which reacts with a gas such as Xenon. An igniter unit, similar to an ignition coil, is used to create the spark. A ballast unit then takes over to maintain the spark and keeps the gas glowing.

What are the advantages of a HID bulb?

_____

LED (L_____ E_____ D_____)

This type of light has no filament. They are very small so in order for the same amount of light to be produced LEDs are usually grouped together in clusters. Electrons move through the diode to produce light. No heat is produced. They light up very quickly.

Give TWO advantages of LEDs:

Research the following types of headlamp and describe the main features that make each different:

- Low beam units/high beam units
- Combined units
- European lens
- American lens
- Projector type

Headlamp aim

Headlamp beams must be correctly aligned so that they give good light. In the UK they must dip to the left, so as not to dazzle oncoming traffic.

Use an optical beam setter to check a vehicle's headlamp aim and dip. Make sure you know what to do. Read the instructions or ask your supervisor. Before starting the check, find out where the adjusting screws are on the headlamps and what each screw alters.

Optical beam setter

Before checking beam alignment make sure that the:

- vehicle is on level ground
- tyre pressures are correct
- suspension and ride height are within manufacturer's specifications
- vehicle is correctly loaded.

VEHICLE LIGHTING SYSTEMS

The vehicle lighting system may be split into two main circuits. These are:

_____

List the basic components required to make these circuits:

_____

_____

_____

_____

_____

_____

_____

How is the earth side of the circuit completed?

_____

The side light circuit normally has two front and two rear lights. There are other lights found at the rear of the car. What are these lights for?

_____

Complete the following side light diagram for a vehicle using a chassis earth return system:

Lighting switch

Front

Rear

The headlight circuit consists of the main and dipped beam. A spring loaded switch is used to operate the flashing signal.

Other circuits which are normally considered part of the main lighting system are:

- _____
- _____
- _____

Stop lamp circuit

Light vehicles, heavy vehicles, coaches and buses are fitted with two rear facing stop lamps. Motor cycles are normally fitted with one rear facing stop lamp. It is also common these days for light vehicles to be fitted with a high level stop lamp in the rear window.

Complete the following stop lamp circuit diagram. Draw and label the fuse and earth connections. The vehicle has a chassis earth return.

Ignition
switch

FUSES

A typical fuse box is shown below.

Typical fuse box

Electrical circuits need to be protected from overload or short circuit that cause excess current to flow. If they are not protected the circuit could heat up sufficiently to melt the insulation and cause a fire. A fusible link or circuit breaker provides protection against such an event.

A fuse box is located in a box or panel, usually found under the dashboard, behind a panel in the foot well or in the engine compartment. The fuses are normally numbered or labelled for easy identification. The fuse box may also contain relays and other control units. More information as to the fuse identification system is available in the owner's or service manual.

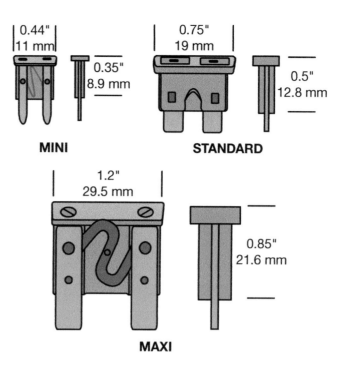

MINI **STANDARD**

MAXI

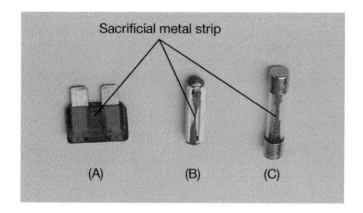

Sacrificial metal strip

(A) (B) (C)

Correctly name the fuses shown above:

A _____

B _____

C _____

There are three basic types of fuses used in motor vehicles. The most common one used today is the blade or spade fuse. This normally comes in three sizes, standard, mini and maxi blade. Older vehicles used the ceramic type of fuse, and some would have used the cartridge fuse. The cartridge fuse can still be found on some applications in vehicles such as radios and other audio equipment.

Fuses are rated by the current at which they 'blow'. Current rating on a blade fuse is indicated by its colour and the rating is also embossed on the coloured plastic casing.

Complete the following table of typical colour coding of blade fuses:

Ampere rating	Housing colour
4	Pink
5	
10	
15	Light blue
20	
25	Natural
30	

Should an excessive current flow in the circuit, due to a fault in that circuit, the fuse is designed to blow and break, protecting that circuit from any damage. The figure beside shows a good and blown fuse.

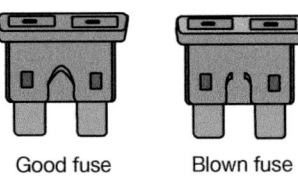

Good fuse   Blown fuse

AUXILIARY SYSTEMS

The modern motor vehicle has a number of electrical systems apart from the starting, charging, ignition and engine management systems. These other systems fall under the heading 'auxiliary systems'.

Windscreen heater

Windscreens can mist up during the cold damp winter weather. This is caused by the condensation and moisture in the air. Some front windscreens and most rear windscreens have a heater element in them.

This is a w_____ that is fitted inside the windscreen. When electrical c_____ passes through the wire it h_____ up and demists the windscreen. A heated windscreen needs a relatively high current to operate effectively.

What is used to prevent the heating element being left switched on for too long?

_____

Immobilizer and alarm

Many car manufacturers now fit some form of immobilizer and alarm to their vehicles. These are designed to prevent or deter thieves from stealing from a vehicle or actually stealing the vehicle itself.

There are two types of systems:

Active anti-theft: These use an _____ to deter any potential thieves.

Passive anti-theft: These use an _____ to prevent the vehicle being started and driven away.

The immobilizer uses a number of vehicle systems to prevent it from being started by a potential thief.

 Light switches and controls should be checked for security and correct operation.

Name FOUR systems on which the immobilizer acts:

1 F_____

2 I_____

3 S_____

4 E_____

Horn

Insert the missing words into the following paragraph using the word bank below (note: there are two extra distracter words):

sound	reduce	flexible	mechanically
transmission	increase	audible	oscillating
volume	body	pneumatically	electrically

All road vehicles must have an _____ warning device that produces a _____ of an appropriate _____. These are known as 'horns' and are normally _____ operated by an _____ (vibrating) diaphragm that is either operated electrically or _____ (by air). Horns should be fitted to the vehicle _____ on a _____ base that will prevent _____ of the oscillations to the vehicle body, which could _____ the effectiveness of the horn.

Windscreen wipers

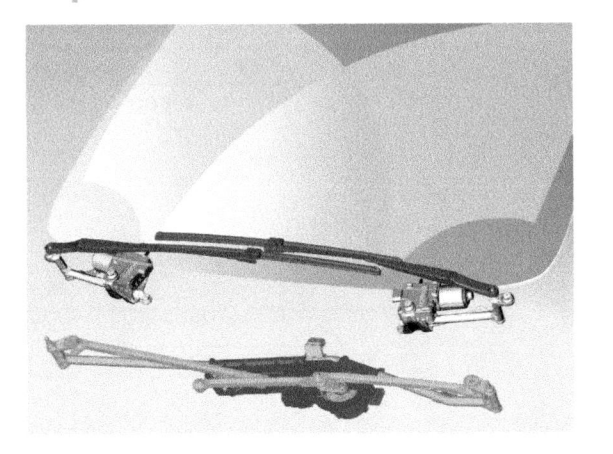

(Top) Two motor system. (Bottom) A typical single motor system

Insert the missing words into the following paragraphs using the word bank below (note: there are two extra distracter words):

cleared	crank	arm	electric
view	windscreens	stop	multi
speed	rains	polarity	power
same	rotating	reverse	move
different	bottom	steering	off
systems	operated	intermittent	after

When it _____ the windscreen needs to be _____ to give the driver a clear _____ of the road ahead. This is achieved by the windscreen wipers. These wiper _____ can be fitted to the front and rear _____. The front windscreen wipers are _____ by a standard _____ motor and _____ arm assembly. As the motor rotates it uses the crank _____ assembly to _____ the wipers back and forth.

It is common for the rear wiper motor to _____ its direction in order for the wiper blade to move back and forth. This is achieved by reversing the _____ (direction) of the _____ current to the motor.

Windscreen wiper motors need to _____ or 'park' in the _____ position when switched _____. This is done by continuing the _____ to the motor _____ the wiper is switched off. As the wipers reach the _____ of the windscreen the power is automatically cut off.

The wiper circuit is normally controlled via a _____ function switch mounted on the _____ column and may have two or three _____ positions and an _____ setting.

RMP Windscreen wipers need to be checked at slow, fast and intermittent operation and for a period when the windscreen washer is operated.

Central locking

Most modern motor vehicles are now fitted with a central locking system. This allows all of the doors, boot and fuel flap to be locked automatically by operating the master door lock. It is quite common for vehicles to have a remote control on the key fob which is used to lock and unlock the vehicle.

When it is operated, the master door lock will activate a m_____ s_____ which will send a signal to the ECU, which in turn sends a signal to the s_____ or door actuators to lock/unlock the vehicle.

A system which is sometimes fitted is the 'l___ l___' system. When the central locking system is activated the electric windows, sunroof and in some cases the convertible hood are closed.

Electric windows

It is very common for vehicles to have electrically operated windows. They are safer and easier to operate, especially the passenger windows. They make use of a DC m____ fitted inside each door. When a t____ position rocker switch is operated by the driver, the motor is operated. Force from the motor to the window glass is achieved by a gearbox and lever mechanism. Some systems use a flexible r____ or c____ to raise and lower the window.

Electric mirrors

Most vehicles now have electric mirrors fitted as standard. These make it safer for the driver as it is an easier task to adjust the mirrors, especially the passenger mirror.

What controls the movement of the mirror glass?

_____

These allow the glass to pivot up/down and left to right. They are normally operated by a joystick type of switch.

In-car entertainment (ICE)

An AM/FM radio with CD and DVD player

Name THREE other systems that ICE includes:

1 _____

2 _____

3 _____

Multiple choice questions

Choose the correct answer from a), b) or c) and place a tick [✓] after your answer.

1 **A battery contains a liquid mixture known as:**

a) Acid [ ]

b) Water [ ]

c) Electrolyte [ ]

2 **The unit of electrical measurement for resistance is the:**

a) Volt [ ]

b) Ohm [ ]

c) Amp [ ]

3 **Which type of lamp is halogen commonly used in?**

a) Indicator [ ]

b) Headlamp [ ]

c) Brake light [ ]

4 **An atom contains a:**

a) Positively charged proton [ ]

b) Positively charged electron [ ]

c) Negatively charged proton [ ]

5 **When disconnecting a battery, which terminal should be disconnected first?**

a) Positive [ ]

b) Negative [ ]

c) Doesn't matter [ ]

PART 6
LOW CARBON TECHNOLOGIES

USE THIS SPACE FOR LEARNER NOTES

_____

_____

_____

_____

_____

_____

SECTION 1

Introduction to low carbon technologies in the automotive industry

USE THIS SPACE FOR LEARNER NOTES

_____

_____

Learning objectives

After studying this section you should be able to:

● Understand how your own actions can affect the environment.
● Understand the impact that a conventional vehicle has on the environment.
● Understand some of the actions vehicle manufacturers are taking to reduce carbon emissions.

Key terms

Carbon footprint   The total emissions caused by an organization, event, product or person.
Carbon monoxide (CO)   Colourless, odourless gas, poisonous to animal life.
Carbon dioxide (CO_2)   Greenhouse gas that contributes to global warming.
Oxides of nitrogen (NO_x)   Can cause respiratory conditions, smog and acid rain.
Sulphur dioxide (SO_2)   Can cause pollution and acid rain.
Soot particles   Cause respiratory problems and cancers.
Hydrocarbons   Cause respiratory problems, liver damage and cancers.
Greenhouse gases   The harmful gases and emissions released into the atmosphere.
Biofuel   Fuel that is obtained from growing crops.
Hybrid   Vehicle using a combination of power sources such as conventional engine and electric motors.
Electric vehicle   Totally reliant on an electric motor for propulsion.

```
A G G C N D S O I C O N E O D N T
P C G L O B A L W A R M I N G N N
S O I S U L P H U R D I O X I D E
C E L D C A R O C B P R O R I R L
E H A L R C Y A I O H U P A O R E
C R Y B U A R N T N S T C D A Y C
A C A B B T I D E M O A L R I R T
N E G O R T I N F O S E D I X O R
C S C R R I C O F N G C R D O T I
E U O A B O D N N O E E E S S A C
R O I A O C O R I X H E E O L R R
S H C A R B O N D I O X I D E I A
S N O B R A C O R D Y H O C U P O
O E L A N O I T N E V N O C F S N
D E C S I O E I C N V C S E O E E
P R R O M C L S O O A I E L I R T
A G O D C R O N R H D R L C B O I
```

CARBONFOOTPRINT
ELECTRIC
HYBRID
CONVENTIONAL
BIOFUELS
GREENHOUSE
CANCERS
LIVER
RESPIRATORY
HYDROCARBONS
SOOT
ACIDRAIN
POLLUTION
SULPHURDIOXIDE
OXIDESOFNITROGEN
GLOBALWARMING
CARBONDIOXIDE
CARBONMONOXIDE

PROTECTING THE ENVIRONMENT

Everything on our planet has influences on our environment. This 'carbon footprint' as it is termed is affected by population, economic output, energy used and how busy the world's economy is. To decrease our carbon footprint everyone must take action, however big or small, to reduce it.

Name other forms of transport that have a smaller carbon footprint than a car.

1 _____

2 _____

3 _____

4 _____

Changing your driving style can have a huge effect on reducing a vehicle's emissions and improving its fuel economy.

In small groups discuss different driving styles that adversely affect vehicle emissions and fuel consumption. List the six styles which you consider to be the most significant below:

1 _____

2 _____

3 _____

4 _____

5 _____

6 _____

Travelling at approx 40 mph in fifth gear uses 25 per cent less fuel than in third gear.

There are many different factors that will affect both the vehicle and the environment. It is important to think and plan ahead. It will only take a small amount of time but can extend the life of the car.

Check tyre pressures

Fill in the missing gaps to complete the following paragraph:

If your tyres are _____-inflated they will create more _____ as they move over the

_____ surface, therefore to _____ your speed the engine has to work _____. This can
increase your fuel consumption by up to 3 per cent.

How often should you check and adjust your tyres?

_____

What else should be checked on a regular basis?

_____

These checks can help your car use less fuel and could increase the life of the tyres, so check
your car manual for the correct pressure.

What adjustments need to be made if the vehicle is going on a long trip with a heavy load?

_____

Clear out any extra weight

Reducing your car's weight and air resistance will help make your car more efficient. Things like
roof racks, bike carriers and roof boxes will affect your car's aerodynamics and reduce fuel
efficiency. Also remove any unwanted items that are in the boot or storage areas, as energy will
be wasted by transporting them around.

Service your vehicle regularly

A well-maintained vehicle tends to be more efficient. Ensure your car is serviced at the correct
intervals according to the owner's manual.

Describe other advantages of having your vehicle serviced regularly:

_____

_____

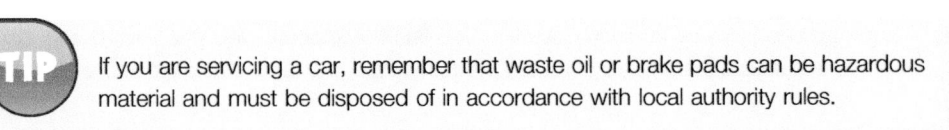

If you are servicing a car, remember that waste oil or brake pads can be hazardous
material and must be disposed of in accordance with local authority rules.

Research what kind of hazardous materials your local authority will dispose of and
how they dispose of them.

Adjusting your driving style could cut your CO_2 emissions by around 8 per cent.
This could add up to an annual fuel saving of up to 1 month per year.

Drive at an sensible speed

The most economical speed to drive at depends upon the vehicle type but is typically around 55
to 65 mph. A higher speed than this can use up to 25% more fuel.

What other advantages will there be from driving at a slower speed?

_____

_____

You should always drive within the national speed limit.

Research how driving at different speeds effects fuel consumption. Draw a graph to indicate how fuel economy (MPG) is effected by different speeds (MPH).

What is the national speed limit for a car or motorcycle on a dual carriageway?

_____

Acceleration and braking

What will be reduced by avoiding sharp acceleration and heavy braking?

_____

_____

Learn to read the road ahead and anticipate what is going to happen, you can then adjust your speed up and down and reduce the amount of braking required.

Research a vehicle's typical stopping distance at the following speeds and complete the diagram below.

Braking distance
Thinking distance

What road conditions will affect the stopping distance of a vehicle?

_____

_____

Using the gears

Changing up gears a little earlier (short shifting) can reduce revs per minute (rpm) and reduce your fuel usage. Changing down a gear at the right time too is important, don't allow the engine to struggle or labour as it will use more fuel.

Apart from using the vehicle's brakes what other ways are there to slow a vehicle down?

_____

What is a modern fuel injection system capable of doing?

_____

_____

When is a vehicle deemed to be under deceleration?

_____

_____

Planning your route

Planning your route in advance can help you take the shortest route and avoid getting lost. It can also avoid congestion and roadworks, thus saving you time, money and wear and tear of the vehicle.

What could aid the planning of a journey?

_____

_____

If you know someone is doing the same journey as you then travelling together in the same car (car sharing) will mean one less car is being used which will reduce carbon emissions.

TIP Avoid driving short journeys as a cold engine uses almost twice as much fuel and catalytic converters can take 5 miles to become effective. Cycle or walk instead if possible.

Using auxiliary equipment

The use of air conditioning should be kept to a minimum.

List other electrical auxiliary equipment used on a vehicle:

_____

_____

Why does using auxiliary equipment increase fuel consumption?

_____

_____

Name the component that charges the vehicle battery? _____

Describe a situation when a vehicle could use most of its auxiliary systems:

_____

TIP Do not rev an engine unnecessarily as it wastes fuel and increases engine wear and emissions.

Apart from the examples already mentioned, how can you reduce your carbon footprint when travelling?

1 _____

2 _____

3 _____

4 _____

5 _____

TIP Careful driving could save you 1 month of fuel over a year.

WWW http://www.direct.gov.uk/en/Environmentandgreenerliving/index.htm

Vehicle fuel emissions

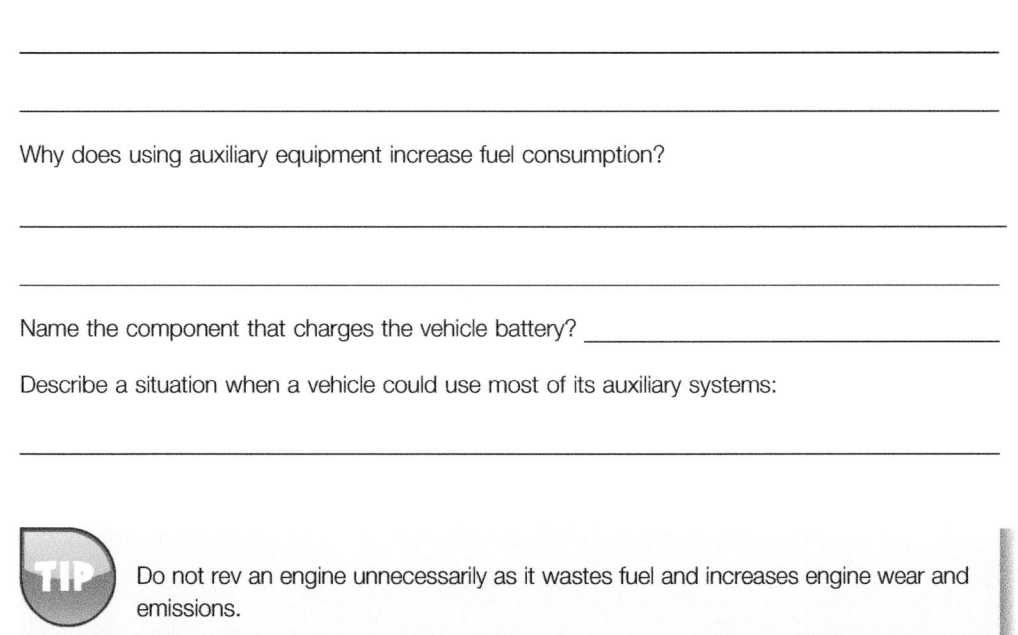

There are three types of automotive emissions. Indicate on the diagram the three types of harmful automotive emissions and where they are emitted from the car?

Explain how a vehicle produces each of the three types of emissions you have identified on page 224.

1 _____

2 _____

3 _____

Below is a list of gases that are produced by a vehicle's engine. Match the chemical symbols to the gases.

Carbon monoxide		CO
Carbon dioxide		NO$_x$
Oxides of nitrogen		SO$_x$
Sulphur dioxide		CO$_2$

What is fitted to a vehicle exhaust system to reduce its emissions? _____

Match the exhaust emissions gases produced by a conventional vehicle with the harmful effects they can have on both humans and the environment in the boxes below.

Carbon monoxide		Pollution and acid rain
Carbon dioxide		Causes respiratory problems, liver damage and cancers
Oxides of nitrogen		Greenhouse gas that contributes to global warming
Sulphur dioxide		Can cause respiratory conditions, smog and acid rain
Soot particles		Causes respiratory problems and cancers
Hydrocarbons		Colourless, odourless, poisonous to animal life

ALTERNATIVE OR GREENER FUELS

Using a car produces carbon dioxide (CO$_2$), this is one of the main greenhouse gases responsible for climate change.

Petrol is the vehicle fuel that creates the most carbon dioxide (CO$_2$). Using a car that runs on an alternative to petrol can reduce the CO$_2$ emissions produced from driving.

Diesel

Diesel engines produce less CO$_2$ than petrol cars but currently produce more pollutants into the air. The emissions of petrol and diesel engines are controlled by European regulations introduced in 1992. For diesel engines this was known as Euro 1.

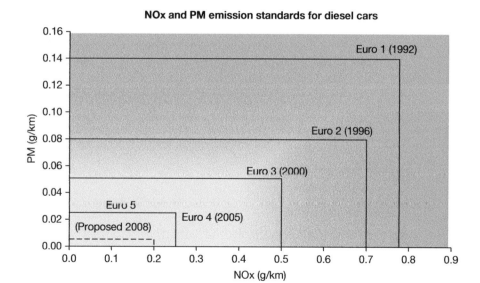

NOx and PM emission standards for diesel cars

From the image shown what is the current European standard for a modern diesel engine?

Alternative fuels

Alternative fuel is a name given to a fuel which is not petrol or diesel and can propel a vehicle. Most of these fuels can be used by a modified conventional petrol or diesel engine.

Liquefied petroleum gas (LPG)

LPG is mainly made up of propane. It is a by-product of oil refining or natural gas. Most vehicles using LPG are 'dual fuel', meaning they can run on either LPG or petrol.

What would be the advantage of having a vehicle that uses both LPG and petrol?

_____

 **TIP** LPG powered vehicles produce about 15 per cent less carbon dioxide (CO_2) than equivalent petrol engines.

One big advantage of using LPG is the cost. LPG costs around half the price of petrol or diesel.

Name THREE other advantages of using LPG:

1 _____

2 _____

3 _____

What are the disadvantages of using LPG:

1 _____

2 _____

Most petrol engines can be converted to use LPG. Conversions have to be undertaken by an approved installer. LPG cars qualify for a lower rate of road tax, as they are classed as 'alternative fuel cars'.

How is the LPG stored on a vehicle? _____

In what form is LPG when it is in the tank? _____

Nitrogen oxide (NO_x) and particulates are both harmful to human health.

Biofuels

Biofuels and biodiesel are alternatives to ordinary petrol or diesel fuel. They are made from plant materials like oil seed rape or wheat which are passed through a refining process and added in varying percentages to ordinary diesel or petrol. This reduces CO_2 emissions.

How do biofuels reduce climate change?

_____

_____

Biofuels can be mixed with ordinary diesel or petrol and used in normal cars. Much of the diesel available in the UK, and some petrol, contains a percentage of biofuel. Petrol contains up to a maximum of 5 per cent and diesel 7 per cent.

What are the main differences between fossil fuels and biofuels?

_____

_____

_____

Read more about fossil fuels and biofuels at
http://www.energyzone.net/aboutenergy/fossil_fuels.asp and
http://www.biofuels.co.uk/

VEHICLE DESIGNS TO REDUCE CARBON EMISSIONS

Modern cars are **not** very energy efficient, despite advances in vehicle technology to improve mpg/lower emissions.

Vehicle manufacturers are designing new ways to reduce CO_2 emissions and improve a vehicle's efficiency. Some of these are identified in the table below. Research the ways manufacturers are trying to improve vehicle efficiency and complete the table. Suggest one other vehicle design used to improve emissions in the final row of the table.

Vehicle design to improve emissions	Definition/how it functions	How it reduces emissions	Manufacturer/car type using this technology
Stop start technology			
Regenerative braking			
Smaller engines			

Name some other ways that vehicle manufacturers are trying to improve vehicle efficiency:

_____

HYBRID AND ELECTRIC VEHICLES

Hybrid cars use a combination of two or more power sources such as conventional engines and electric motors to improve energy consumption.

Electric vehicles are 100% battery powered and in ideal conditions can travel approximately 100 miles between charges. The numbers of electric vehicles are expected to increase considerably in the future.

Fill in the missing words in the following paragraph using the terms below (note: there are two extra distracter words):

petrol	**electric**	**vehicles**	**low-speed driving**
emissions	**power**	**braking**	**accelerating**
mains power	**recharges**	**diesel**	**microprocessor**

Hybrid vehicles use a small _____ or _____ engine combined with an

_____ motor which provides additional _____ to assist the engine in

_____, passing, or hill climbing. Some _____ use only the electric motor

for _____ reducing the vehicles' _____ significantly. The battery is

controlled by a _____ and _____ as you drive.

At slower speeds (up to 30 mph), the vehicle runs on battery power. This uses less petrol/diesel when driving and reduces CO_2 emissions.

Name the advantages and disadvantages of a hybrid over an electric vehicle?

Advantages:

1 _____

2 _____

Disadvantages:

1 _____

2 _____

Name the advantages and disadvantages of a hybrid or electric vehicle over a petrol/diesel powered one.

Advantages: _____

Disadvantages: _____

_____

http://www.fueleconomy.gov/feg/hybridtech.shtml

In groups, research the different types of hybrid cars available on the market and identify which has the best MPG.

Complete the table below to show the different percentages of emissions produced by modern vehicles.

Fuel type	CO_2	CO	PM	NO_x	HC	Result
Petrol						
Diesel						
LPG						
Hybrid						
Electric						

TIP Hydrogen-powered vehicles are also available. They produce zero emissions but are not easily available.

Name one other disadvantage of hydrogen-powered cars.

_____

RECYCLING

When a car comes to the end of its life, it must be taken to an authorized End of Life Vehicle recycler (ELVs) or Authorized Treatment Facilities. The aim of this is to safely recycle as much as possible and to reduce the amount of waste from each car when it is finally scrapped.

Name the common vehicle parts that may be recycled:

1 _____

2 _____

3 _____

4 _____

5 _____

6 _____

7 _____

8 _____

In small groups discuss the advantages of recycling.

www http://www.scrapcars.com/resources/18.html

Multiple choice questions

Choose the correct answer from a), b) or c) and place a tick [✓] after your answer.

1 **What does the term carcinogenic mean?**

   a) Induces the risk of cancer [ ]

   b) Is a type of biofuel [ ]

   c) Is a liquid that absorbs water [ ]

2 **Which of the following modes of transport have the smallest carbon footprint?**

   a) Walking and driving [ ]

   b) Walking and biking [ ]

   c) Walking and catching the train [ ]

3 **A hybrid vehicle has a:**

   a) Large engine and electric motor [ ]

   b) Large batteries only [ ]

   c) Small engine and electric motor [ ]

4 **Name the most common gas in the atmosphere:**

   a) Nitrogen [ ]

   b) Oxygen [ ]

   c) Carbon dioxide [ ]

5 **What health problem may be caused by breathing dirty or contaminated air?**

   a) Asthma [ ]

   b) Eozoma [ ]

   c) Dermatitis [ ]

PART 7
VALETING

USE THIS SPACE FOR LEARNER NOTES

SECTION 1
Basic vehicle valeting 231

Basic vehicle valeting

USE THIS SPACE FOR LEARNER NOTES

_____

_____

_____

_____

_____

Learning outcomes

After studying this section you should be able to:

● Identify commonly used valeting tools and equipment and understand how they are used correctly and safely.

● Identify commonly used cleaning materials and understand how they are used correctly and safely.

● Understand the correct procedures for valeting motor vehicles safely and effectively.

Key terms

Engine laquer   A fast-drying clear acrylic varnish which enhances the appearance of the engine and its components. The varnish protects the vehicle from moisture and corrosion.

Rubber dressing   Cleans and restore tyres, rubber trims and interior plastics to an 'as new' appearance.

Cutting compound   A paste used for removing scratches, blending in new paintwork and removing paint over-spray.

Chamois Automotive   Drying cloth that is safe on acrylic, lacquer, enamel, and polyurethane paints and clear-coats.

Key valeting equipment

Bucket
Sponge or wash mitt
Soft body brush
Wheel cleaner
Wheel brush or dish brush
Car shampoo or wash '&' wax
Car polish and car wax
Application sponges
Microfibre cloths
Glass cleaner
Spray foam interior cleaner
Various scrubbing brushes
Tyre dressing
Vacuum

```
D U S C U R R T I V M U E O W L
H C M E A C T S K M S W A L S R
F R Y A E G S H R C B I M O E P
M C L W W H I T H B P I E N O V
E P B B A H U Y E T U U A E H S
T S M M I X E R E H B E E P R R
M P P G N C H E T U L V A M I O
A O W O L E L D L C W A I E P P
O L F U N A L R L B U C K E T U
F I R R I G S E P C R U P T L A
Y S M N Y E E S A O W U I M S S
A H M N V H S S F V N M S S E E
R A A S W D I I B G L S L H M S
P G E R G I B N H H O S E N C S
S E H S U R B G N I B B U R C S
C A E T E G K I R X B N O I S F
```

BUCKET
SCRUBBINGBRUSHES
SPRAYFOAM
TYREDRESSING
VACUUM
GLASS
MICROFIBRE
SPONGES
WAX
POLISH
SHAMPOO
WHEELBRUSH
WHEELCLEANER
MITT

It is essential to choose the correct products to carry out valeting operations quickly and effectively, but also SAFELY. The regulation, Control of Substances Hazardous to Health (COSHH), requires manufacturers of valeting products to make detailed information available on the product container.

For more specific information and guidance visit
http://www.hse.gov.uk/mvr/priorities/degreaser.htm

Different products need to be used with care, especially as they could contain a 'cocktail' of possibly hazardous chemicals if mixed.

VALETING

Professional valeting today requires far more than just a simple bucket and sponge, although they still have their uses. There is now a vast range of specialist preparations, each designed to do a particular job to clean a vehicle in the most efficient way possible.

TOOLS AND EQUIPMENT

Complete the table shown below which lists the different types of cleaning products available. Identify where these products will be used and suggest two others.

Product	Typical components where used

Before using any valeting product, where can basic information regarding their proper use and appropriate precautions be obtained?

_____

Below is a typical COSHH data sheet. List all of the dangers and hazards associated with heavy duty fabric cleaner.

Product Safety Data Sheet

Product Code: 214

Product: HEAVY DUTY FABRIC CLEANER

SECTION 1: PRODUCT DESCRIPTION & PHYSICO CHEMICAL DATA

APPLICATION: Detergent Concentrate for use through hot water soil extraction carpet/fabric seat cleaning machines.

PHYSICAL FORM: Water thin clear blue liquid, citrus odour.

CHEMICAL COMPOSITION: Sequestering agents, surface active agents, sodium hydroxide, 1-Methoxypropan-2-ol, hydrotope, optical brightener, Perfume oils, dye, preservatives, water.

HAZARDOUS INGREDIENTS:

Material	Nature of Hazard
Sodium Hydroxide	Causes damage to eyes, skin and clothes.
1-Methoxypropan-2-ol	Maximum exposure limit set (M.E.L.) (see below).
	Can be absorbed through skin.
M.E.L. = 1-Methoxypropan-2-ol =	8 hr TWA: 100 ppm
	10 min TWA: not set

SECTION 2: FIRE/EXPLOSION/REACTIVITY DATA

Product is non-flammable, However, irritating fumes may be given off in the event of a fire. Treat fires with dry chemical, foam or waterspray.

Reactivity: Alkaline products react with light metals (such as aluminium, tin or zinc), with the evolution of hydrogen gas which is highly flammable.

SECTION 3: HEALTH HAZARD INFORMATION

Personal protection/precautions in use. Use only according to instructions. Always wear impervious gloves and eye protection when handling this chemical. Do not mix with other chemicals. Always maintain a good standard of occupational hygine when handling chemicals.

Effects of over exposure

Skin contact: Degreases skin, may lead to redness and cracking. Irritation will be felt on sensitive skin. Prolonged contact causes damage.

First Aid: Remove contaminated clothing, wash affected area with soap and water.

Eye contact: Severe pain, redness watering. May cause permanent damage.

First Aid: Wash chemical out of eyes immediately. Use plenty of clean water for at least 15 minutes. Seek immediate medical attention.

Ingestion: Sore mouth and throat, abdominal pain, vomiting.

First Aid: Rinse mouth out with water then drink milk or water and seek immediate medical attention.

Inhalation: Mild irritation.

First Aid: Move casualty to fresh air, seek medical attention if necessary.

SECTION 4: STORAGE

Store away from children. Always ensure cap is tightened after use. Keep away from food stuffs.

MAX/MIN TEMPERATURE: 0°C–35°C

SECTION 5: SPILLAGE/LEAK PROCEDURE/WASTE DISPOSAL

Small spillages can be rinsed away to drain. Large spillages should be contained and advice sought from Local water company.

Disposal: According to local regulations.

Name one material that heavy duty fabric cleaner reacts with. _____

How should a fire involving this product be treated?

_____

How should small spillages be dealt with? _____

It is important to always wear suitable PPE for each individual task.

List FOUR forms of PPE that should be worn by anyone cleaning vehicles.

Body part to be protected	PPE	When required

Name the THREE main areas of a vehicle that a valet would work on:

1 _____

2 _____

3 _____

EXTERIOR VALETING

An exterior valet is an excellent way of reinstating and improving the overall look of a vehicle. However, on an older vehicle the valeter's job will require a number of steps to show the best results.

What will help protect a vehicle's paintwork?

_____

TIP Do not wash your car on an excessively hot or sunny day. The water will evaporate too quickly and you will get soap marks on your car.

In small groups identify the equipment and products that would be required for a full **exterior** valet.

- _____
- _____
- _____
- _____
- _____
- _____
- _____
- _____
- _____
- _____
- _____
- _____

Name the main benefits of washing a vehicle by hand:

_____

_____

Name the TWO systems of washing a vehicle commonly used in the motor trade:

- _____

_____

- _____

With both systems, best results are obtained by using plenty of water. Use a gentle sweeping action for the pressure washer and small circular movements for the sponge.

Cleaning action

To obtain the best results you should:

Apply soap and water mix starting at _____

Rinse with clean water, starting at _____

To protect a vehicle's paintwork from scratching before and during washing, what should a valeter undertake on a regular basis?

_____

_____

_____

A modern sponge is made from basic cellulose material, although real sponges can still be used. A mitt is perhaps less likely to cause scratches than a sponge, since mitts are made of smooth, soft synthetic fibres.

TIP Many different body shampoos are now available, always read the instructions before use and never exceed the prescribed mixing requirements as this could cause paint damage.

Pressure washer

Place the items listed below next to their relevant locations indicated on the diagram.

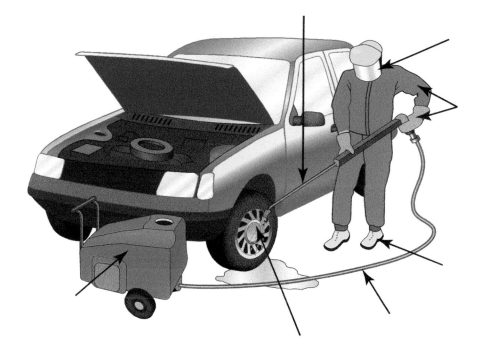

High pressure spray

Portable pressure washer

High pressure hose

Rubber boots

Protective clothing/gloves

Face mask

Lance

The use of hot water/steam pressure cleaners will require training before use.

Name the THREE main sources of potential danger when using a pressure washer:

1 _____

2 _____

3 _____

Drying a vehicle

It is important to dry a vehicle after cleaning it as the impurities in the water or remnants of shampoo can leave marks or spots on the paintwork or windows.

Complete the following paragraph:

Drying can be undertaken using a _____ leather or _____ cloths. These cloths are used because they are designed especially for this operation. They are _____ at _____ than any other sort of cloth. You may need more than one and they must be _____ out regularly.

 TIP Professional valeters often use a squeegee for speed prior to finishing with chamois or cloth.

Apart from the main exterior panels of the body, what other areas should a valeter make sure are clean and dry?

_____

 TIP Cleaning chemicals and detergents can be harmful to the environment. Make sure that the area used for washing a vehicle meets all local environmental regulations. It is particularly important to prevent contaminants from flowing into water drainage systems.

Tar removal

Tar is transferred to the vehicle's body as the tyres pass over the tarmac surface. When is this most likely to occur?

_____

To remove tar, a special solvent-based product can be used. It is most important that the product removes only the tar that is adhered to the body without affecting or damaging the vehicle's paintwork.

What vital features must be observed when using this product?

_____

Wheels and tyres

Most modern cars have steel wheels with wheel trims which are made of either plastic or resin. Others have aluminium alloy wheels.

Road wheels, particularly the front ones, tend to get dirtier more quickly. Explain why this is.

_____

_____

TIP When cleaning wheels, wetting them and leaving the water to loosen the dirt will achieve the best results.

Complete the paragraph below using a selection of words from the word bank (note: there are two extra distracter words):

product	dressing	rinse	difficult	soap
minutes	shiny	dust	dirt	mop

Spray the cleaning _____ onto wheels and leave for a few _____, work the product in to the

_____ to reach areas with a tyre brush (more than one type might be needed) and then

_____ off. It removes brake _____ and road _____ quickly, easily and safely leaving an excellent _____ finish. To add a finishing touch to the task, tyre _____ can be applied.

TIP Do not leave wheel cleaning products on the wheels too long as damage can occur, particularly on alloy wheels. Always read manufacturers' instructions first.

What precautions must be taken when applying tyre dressing?

_____

Polishing and buffing

Before polishing can begin, the surfaces must be thoroughly clean and dry. Buffing polishes cuts away the road deposits on the surface of the old paint to restore its colour and shine.

Why do the abrasive qualities of different polishes vary?

_____

It is important to realize that some paints are harder (more abrasive resistant) than others, so you have to be careful when using these products.

Complete the paragraph below:

Polishing can be undertaken by _____ or by using a _____. The process is the same. Apply

the _____ to a suitable _____ type material and using a very light _____ rubbing action

apply in _____ areas or as directed. Do not rub _____ as this can mark the _____ surface;

also ensure that each action _____ the next.

TIP There are many different types of polish, always adhere to the manufacturers' instructions for application and removal.

When buffing or polishing a car, why is it necessary to ensure that each circular action overlaps the next?

_____

The polisher shown below uses two types of mop heads. They are a sponge and a soft pad.

A polyurethane foam (sponge) is used to: _____

A soft pad is used to: _____

 When buffing a vehicle always ensure that a clean cloth or pad is used. Any dirt on these can cause small scratches in the surface of the paint.

What precautions should be noted when using any electrical tool?

1 Supply must be via: _____

2 Cable and plug condition: _____

3 Cable routing: _____

 Electrical risks can be minimized by using a 110 V electrical supply.

Both during polish application and buffing, the direction of hand movement or polishing tool is important. See the sketch below and label it to indicate which action shows polish application and which shows buffing off.

 Polishes used for this purpose can have a mildly abrasive or cutting action.

 Investigate the different types of products used to cut and polish vehicles and list your findings.

_____

What cleaning product is used to clean a vehicle's chrome parts?

_____

INTERIOR VALETING

Vacuum cleaning

Where should mats and a vehicle's personal effects be stored?

_____

What area of a vehicle usually gets the most use?

_____

Move the seats backwards to expose the carpet fully. Moving the vacuum vigorously across the upholstery and carpets moves the dirt to the surface.

Where should you start vacuuming?

_____

You can use a very stiff brush on upholstery seats and carpet, be quite vigorous for best results.

Cleaning carpets and upholstery

Slightly dirty fabric can be cleaned by hand using a dry foam fabric and upholstery cleaner.

Name THREE advantages of using this type of fabric cleaner:

1 _____

2 _____

3 _____

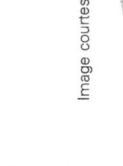 **TIP** If you are using a cleaning agent, test it in a small inconspicuous area. This is to check that it does not harm the material.

On heavily marked fabrics you may have to use a spray/extraction machine. This is a hot water/detergent spray combined with a vacuum extractor to remove the moisture and the dirt as seen below.

Image courtesy of Draper Tools Ltd.

TIP Leather must NOT be treated with petroleum-based solvents, detergents or caustic soaps, otherwise permanent (and expensive) damage can occur. Leather may be cleaned by washing with warm water (not too much), and special soap or cleaner.

Windows

With the windows cleaned on the outside the inside can now be done. A damp microfibre cloth works well on windows or a window cleaner product can be used.

What are the TWO advantages of using a specialist window cleaner product?

1 _____

2 _____

TIP Don't forget to lower all opening windows by about 2 to 5 cm to allow for cleaning the top edge and for ventilation. Be careful not to wet any electrical components.

Plastics

When cleaning dashboards or other vinyl surfaces, why should you always spray your product cleaner on to the cloth, not directly on to the plastic?

_____

_____

_____

If you apply too much product it can run down into the cracks and difficult-to-reach areas which will attract dust and dirt.

TIP Use a paintbrush as a duster while cleaning air vents, dashboard and around the centre consul. Wrap some tape around the metal part to prevent scratches.

Name the types of waste materials that would be generated when a vehicle is being valeted:

_____

_____

It is most important that the disposal of waste, inevitable with the valeting process, is dealt with safely, in accordance with statutory and organizational requirements.

Why is it important in any valeting operation to use only the recommended proportion of solvents or soaps, even if the vehicle is particularly dirty?

_____

_____

Complete the table below. Some examples are shown for you.

Where on vehicle	Vehicle exterior	Alloy wheels	Body work	Glass & mirrors	Plastic trim	Tyres
Product type		Wheel cleaner				
Preparation required						
How to use it		Brush/spray on. Pressure wash off				
Safety	Irritant					

Multiple choice questions

Choose the correct answer from a), b) or c) and place a tick [✓] after your answer.

1 What does COSHH stand for?

a) Control of Systems Hazardous to Health [ ]

b) Control of Substances Hazardous to Health [ ]

c) Care of Substances Hazardous to Health [ ]

2 What does EPA stand for?

a) Environmental Protection Act [ ]

b) Environmental Procedure Act [ ]

c) Earth Protection Act [ ]

3 Before a vehicle is valeted what must you ensure?

a) Handbrake is applied and ignition keys removed [ ]

b) Rear wheels are chocked and battery lead removed [ ]

c) Steering lock is activated and the vehicle is alarmed [ ]

4 What precautions must be observed when using aerosols?

a) Use in a confined area [ ]

b) Use in a well-ventilated area [ ]

c) Use only when the job is completed [ ]

5 Where must any valuables found in a vehicle be stored?

a) In the glove box [ ]

b) In a secure and locked area [ ]

c) In the locked boot area [ ]

PART 8
HEAVY VEHICLE MAINTENANCE

USE THIS SPACE FOR LEARNER NOTES

_____

_____

_____

_____

_____

_____

_____

SECTION 1

Heavy vehicles

USE THIS SPACE FOR LEARNER NOTES

_____

_____

_____

_____

_____

_____

_____

_____

Learning objectives

After studying this section you should be able to:

● **Understand the main differences between light and heavy vehicles.**
● **Identify the main layouts of heavy vehicles.**

Key terms

Heavy Goods Vehicles (HGV) or Large Goods Vehicles (LGV) Any truck weighing over 3500 kg according to European Union regulations.

Public Service Vehicles (PSV) or Passenger Carrying Vehicles (PCV) A vehicle that carries more than eight passengers for a fare e.g. bus or coach.

Service brake The primary braking system. This brake is typically operated by foot and is mechanically separated from the parking brake or emergency braking system.

Secondary brake The emergency brake. This is not powered by hydraulics and is independent of the service brakes used to slow and stop vehicles.

HGVs, LGVs, PSVs and PCVs

Heavy Goods Vehicles (HGV) or **Large Goods Vehicles (LGV)** have always been an integral part of the motor vehicle industry. Their ability to transport both small and very large loads to different locations and countries continues to be their main role.

The main differences between a HGV and a light vehicle (LV) are not that many. Engines, gearboxes, suspension and drivelines are the same in both types of vehicle but the size and strength alters according to the operational requirements of a specific vehicle. A HGV is classed as any vehicle over 3500 kg with a few exceptions.

Public Service Vehicles (PSV) or **Passenger Carrying Vehicles (PCV)** also share many of the same major components as the HGV. They do have additional safety requirements which are different from that of a HGV as they have to ensure passenger comfort and safety as well.

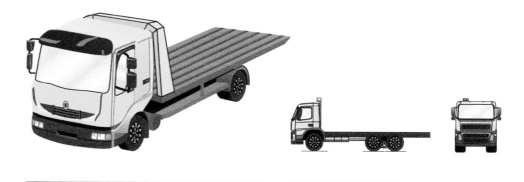

VEHICLE LAYOUTS

HGVs and PSVs have a fairly standard vehicle layout.

Name each of the vehicles that follow. Identify on the diagrams below where the engine is located and which is the driven axle.

From the diagram below name all of the different drive combinations that can be used on this type of HGV.

_____

How can wear and tear be reduced on many modern HGVs?

_____

_____

_____

What other advantage would this have?

_____

_____

Name the device that is used to connect a tractor unit to a trailer.

_____

_____

BRAKES

The only major difference between LV and HGV or PSV brakes is the medium that is used to transfer the force applied by the driver to the braking system. Some vans use brake fluid, others use air and fluid and all large HGVs and PSVs use air.

R	E	Y	R	D	R	I	A	E	O	I	U	AIRDRYER
V	F	O	O	T	V	A	L	V	E	S	N	SPRINGBRAKE
N	A	N	S	U	E	V	R	R	O	A	L	DIAPHRAGM
D	S	Y	S	L	U	S	R	P	T	V	O	SECONDARY
S	I	R	E	S	E	R	V	O	I	R	A	SERVICE
E	K	A	R	B	G	N	I	R	P	S	D	COMPRESSOR
R	E	D	P	E	P	P	S	D	S	O	E	UNLOADER
V	D	N	M	H	B	I	B	V	S	A	R	RESERVOIR
I	L	O	O	P	R	M	P	S	A	H	C	PIPES
C	R	C	C	I	I	A	A	E	R	L	R	CHAMBER
E	M	E	I	O	D	A	G	H	S	R	L	FOOTVALVE
V	H	S	O	E	B	C	V	M	C	V	O	

The diagram below shows a basic air brake system. Label the diagram using the components listed in the word bank below.

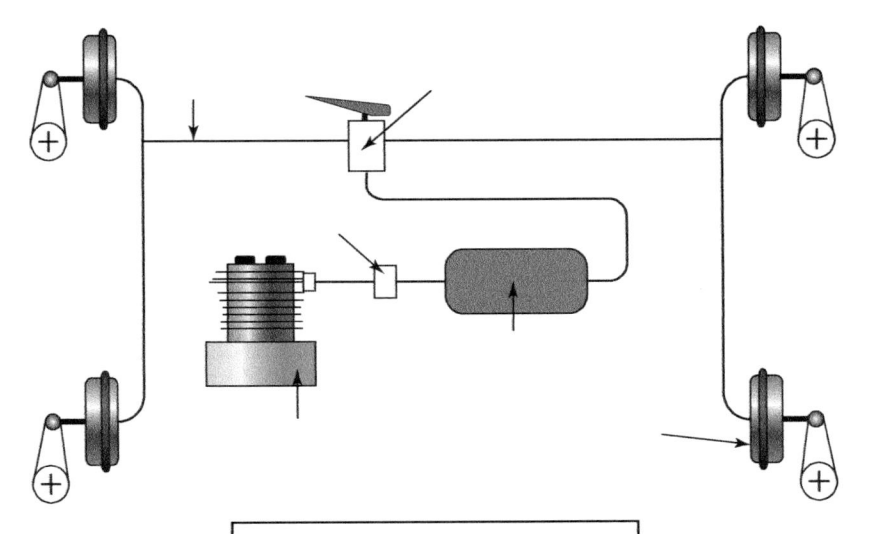

Pipes	Reservoir
Driver's foot valve	Brake chambers at the wheels
Unloader valve	
Compressor	

In small groups research and discuss the use and purpose of the components listed at the bottom of page 244.

Some air brake systems can still be quite basic but most modern systems incorporate the use of an electronic braking system (EBS) working in conjunction with the air.

What is the advantage of using electronics in a braking system?

_____

_____

A HGV or PSV uses two separate braking systems, service brake (foot brake) and secondary brake (emergency). This is to ensure that the vehicle will always have one or more system available in case of emergency.

Research the **minimum** braking efficiencies required for an HGV of PSV to pass an annual MOT test:

_____   _____   _____

http://www.dft.gov.uk/vosa/publications/manualsandguides/
vehicletestingmanualsandguides.htm

Endurance brakes

This is the name of a device that is used to reduce the speed of a vehicle without having to use the vehicle's main or foundation brakes. These tend to fall into two main categories, engine brake and transmission brakes.

Exhaust brake

An exhaust brake restricts the flow of exhaust gases exiting the engine. It works by closing a butterfly valve located in the exhaust manifold. This generates a high pressure in the exhaust manifold and each of the engine cylinders turn into small compressors and act as a brake against the engine rotating. This then slows the vehicle.

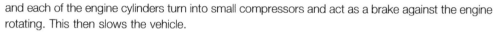

Intarder

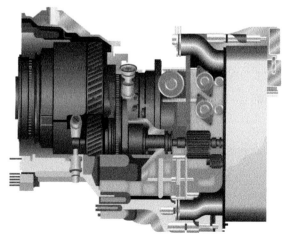

An intarder is part of a vehicle's gearbox. It uses high pressure fluid to act on moving parts in a gearbox and slow them down. This offers resistance to the transmission and the vehicle is slowed.

Retarder

These are electrically controlled transmission brakes. They work by activating large electromagnetics around a spinning shaft, usually the prop shaft. The driver can control how strong the magnets become by increasing or decreasing the amount of electricity allowed into the retarder. The stronger the magnets the greater the resistance on the spinning shaft.

LEGISLATION

HGV and PSV are covered by far more complex legislation than light vehicles. List some of the sections of legislation where HGV/PSV differ from light vehicles. The first two are completed for you:

1 Driving licences.

2 Weight restrictions.

3 _____

4 _____

5 _____

To operate a HGV or PSV you need an operator's ('O') licence. This is a legal requirement that has many parts to it. If you do not abide by all parts, action will be taken by the traffic commissioner and you could have your licence revoked. All vehicles above 3.5 tonnes gross vehicle weight (gvw) that are used to carry goods must have one.

LEGAL OBLIGATIONS

Roadworthiness inspection

Driver daily walk round checks

List the items that you think should be checked on this kind of inspection:

_____

Inspection of all items covered on the statutory annual test

Where would an operator find the information for this kind of inspection?

_____

Preventative maintenance inspection (PMI)

This kind of inspection is generally undertaken every 6 weeks, this is the norm. The timeframe can vary depending on the usage of the vehicles and the conditions in which they operate. The manufacturers' guidelines also have to be adhered to.

When carrying out a PMI what must a technician ensure he completes to comply with the 'O' licence requirements?

_____

TIP As part of your 'O' licence obligations, and the Road Traffic Act 1988, your declaration to ensure your vehicles are maintained to a roadworthy standard means you have agreed to undertake regular inspections. This undertaking includes your drivers' daily walk round checks and preventative maintenance inspections.

PCV roadworthiness and maintenance inspections

In addition to the PMI and the driver daily walk round check, passenger carrying vehicles must adhere to the build standard of the certificate of initial fitness provided when they were newly supplied.

TIP Any vehicle that exceeds 3 500 kg should be taken for its initial MOT test before the end of the first anniversary month of its date of registration, unlike a light vehicle which is 3 years.

How long are records kept of all safety inspections, routine maintenance and repairs?

10 months   12 months   15 months

_____

What do drivers use to record their driving hours?

_____

WWW http://www.fta.co.uk/index.html

Lifting axles

Some HGVs are fitted with **lifting axles** that can be mechanically raised or lowered. The axle is lowered to increase the weight capacity, or to distribute the weight of the HGV's load over more wheels. There are legal requirements in regard to number of axles and a vehicle's weight which are shown in the table opposite.

VEHICLE TYPE AND NUMBER OF AXLES	MAX WEIGHT LIMITS (kg)
Rigid motor vehicles	
2	17 000
3	25 000 (26 000 with road friendly suspension)
4 or more	30 000 (32 000 with road friendly suspension)
Articulated vehicles	
3	25 000 (26 000 with road friendly suspension)
4	32 520 (35 000 with road friendly suspension)
5	38 000
	-44 0002
6	44 0002
Drawbar combinations	
4	32 520 (35 000 with road friendly suspension)
5	32 520 (38 000 with road friendly suspension)
6	44 0002

NATIONAL SPEED LIMITS

Goods vehicles with a design weight over 3.5 tonnes and buses with more than eight passenger seats (regardless of weight) registered on or after 1 January 2005, are required to be fitted with a road speed limiter. The limiter will restrict the maximum powered speed to 56 mph (90 km/h) for goods vehicles, and 62 mph (100 km/h) for buses.

What is the maximum speed that a 44 tonne truck can travel on a dual carriageway?

_____

What is another restriction that applies to HGVs on motorways?

_____

Multiple choice questions

Choose the correct answer from a), b) or c) and place a tick [✓] after your answer.

1 **What type of suspension system is classed as "Road friendly"?**

a) Leaf springs [ ]

b) Coil springs [ ]

c) Air bags [ ]

2 **What does an articulated truck use to connect the air and electrics to the trailer?**

a) Steel pipes [ ]

b) Flexible plastic hoses [ ]

c) Rubber hoses [ ]

3 **Name the combination of braking systems that some vehicles use.**

a) Air and hydraulic [ ]

b) Air and coolant [ ]

c) Hydraulic and coolant [ ]

4 **What does a brake system use to apply the parking brakes?**

a) Air [ ]

b) Fluid [ ]

c) Springs [ ]

5 **What do modern PSVs use to allow for easier and safer passenger access?**

a) The vehicle to kneel [ ]

b) Better lighting [ ]

c) Softer tyres [ ]

PART 9 MOTORCYCLE MAINTENANCE

USE THIS SPACE FOR LEARNER NOTES

SECTION 1
Motorcycle maintenance 250

SECTION 1

Motorcycle maintenance

USE THIS SPACE FOR LEARNER NOTES

_____

_____

_____

_____

_____

_____

Learning objectives

After studying this section you should be able to:

● **Identify the main motorcycle and moped components that require maintenance.**

● **Describe the basic motorcycle maintenance procedures.**

Key terms

Alternator   Charges the battery and provides electrical current for the motorcycle electrical systems.

Chains   Used to transmit power to the motorcycle wheels.

DRLs   Daytime Running Lights.

Handlebars   Used for steering a motorcycle or moped. This is also a typical mounting place for the vehicle's controls.

Moped   A motorcycle that weighs less than 250 kg and has a maximum design speed not greater than 30 mph.

Motorcycle   A vehicle having less than four wheels and weighing less than 410 kg unladen.

Sprocket   A profiled wheel with teeth.

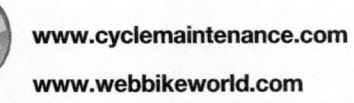

www.cyclemaintenance.com

www.webbikeworld.com

A motorcycle is a vehicle having less than four wheels and weighing less than 410 kg unladen.

A moped is a motorcycle that weighs less than 250 kg and has a maximum design speed not greater than 30 mph. With an internal combustion engine it must have a capacity not exceeding 50 cm³. The moped does not require pedals to propel it.

It is important to follow the manufacturer's maintenance schedule, replacing and maintaining those components/systems as stated. All defects need to be correctly recorded and the owner must be notified of any concerns immediately. Complete all maintenance records.

Identify the tools and equipment that would be used during routine motorcycle maintenance. Search on the Internet and look for tools in your workshop.

Discuss with your tutor what tools are needed and used.

There are a number of motorcycle systems which need to be regularly maintained. Correctly label the following components on the figure below:

Headlamp
Seat
Fairing
Exhaust system
Front tyre
Rear brake disc
Front suspension
Mudguard
Engine
Chains
Sprockets

Front brake assembly

Insert the missing words into the following paragraphs (note: there are two extra distracter words):

topped	manually	low	pads	rod
handlebars	atmosphere	replacing	reservoirs	
two	new	cable	topped	
four	specification	sealed	water	
thickness	old	seat	thin	

Motorcycles usually have _____ brake fluid _____, one for the front brake which is

normally located on the _____, and one for the rear brake which can be tucked away

somewhere under the _____. These should both be checked and _____ up if necessary.

Topping up _____ brake fluid should only be done using _____ fluid from a

_____ container as it tends to absorb _____ from the _____. The fluid also

needs to be of the correct _____.

Before topping up the brake fluid, check the _____ of the brake _____. If they are worn too _____ they will need _____ before the fluid is _____ up.

Some smaller and older motorcycles may use a combination of _____ and _____ to operate the brakes. These will normally need to be adjusted _____.

Why is it important not to spill brake fluid on to painted surfaces of a motorcycle or moped?

_____

STEERING AND SUSPENSION

Insert the missing words into the following paragraphs (note: there is one extra distracter word):

security	maintenance	excessive	movement
colour	lubricated	wear	leaks

The steering and suspension system needs to be checked during routine _____

Check the steering for _____ and excessive _____ and the suspension dampers for _____ and free _____. All pivot points need to be checked for _____ play.

Where necessary the pivot points should be _____.

WHEELS AND TYRES

The only contact a motorcycle has with the road is via the very small area of tread beneath each tyre which is actually touching the road surface. Consequently, the correct size and type of tyres and their condition are vital to rider safety.

List THREE basic tyre maintenance checks that should be carried out on tyres:

1 _____

2 _____

3 _____

Bridgestone

Wheels also need to be checked for:

1 _____

2 _____

3 _____

4 _____

ENGINES

 Wear suitable gloves to protect your skin from the engine oil.

Routine servicing and maintenance of the engine is important as this helps to prolong the life of the engine, improve fuel consumption and control emissions.

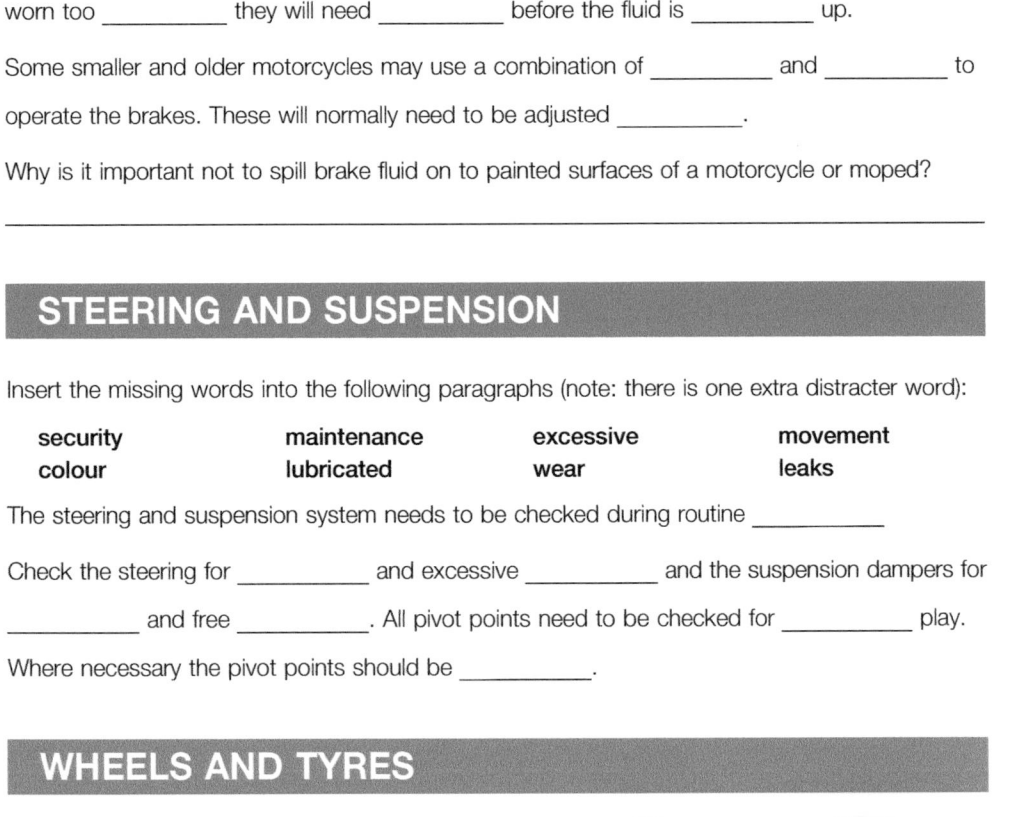

Motorcycle engine

Oil and filter change

Before draining engine oil the motorcycle needs to be on a flat level surface.

Why is this? _____

When draining engine oil, the engine needs to be warm. Why is this?

_____

Drain the oil into a container. Always remember to select and use the correct hand tools to undo the drain plug and oil filter.

When refilling the engine oil make sure you use the correct specification and quantity.

Comma Oil

Motorcycle engine oil

COOLING SYSTEM

A motorcycle engine is not normally enclosed; the passage of air over the cylinders when the cycle is in motion provides an adequate cooling flow.

Liquid cooling

The cooling system is very important as it maintains the correct engine temperature, preventing it from overheating and causing severe engine damage.

List THREE basic checks that need to be routinely carried out on a motorcycle cooling system:

1 _____

2 _____

3 _____

Air cooling

Some motorcycle engines are air cooled and require very little maintenance. When the motorcycle is in motion the air passes over the fins and removes the heat from the engine.

Two examples of motorcycle cylinder heads and barrels are shown below. Label each with the engine cycle on which they would work:

_____   _____

What regular maintenance should be carried out on cooled engines?

_____

Transmission

Most motorcycles use the engine oil to lubricate the transmission.

The chain and sprockets need to be routinely checked for wear and if excessively worn they must be replaced.

What visual checks need to be made of the sprocket?

_____

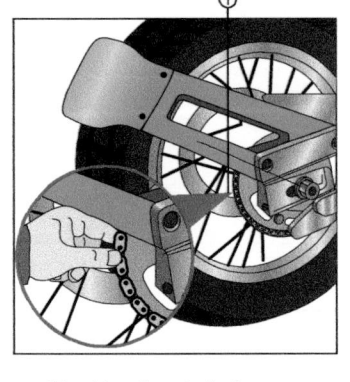

Checking the chain for wear

Briefly describe the check shown above: _____

_____

Clutch fluid

This is the same fluid as used for the braking system. The same procedures and precautions must be observed.

What basic routine check must be carried out with the clutch fluid? _____

ELECTRICAL SYSTEMS

Horn

This is fitted to give an audible warning to other road users and pedestrians. It is normally operated by a push button on the handlebars.

What TWO checks need to be carried out on a horn?

1 _____

2 _____

Lighting

All lamps and reflectors must be kept clean and in good working order. Where the lamp is red or amber no white light must be showing, e.g. broken lens.

Complete the following table by drawing lines to the correct colour for each lamp:

Lighting equipment and reflectors	Colour
Headlamp	Red
Tail lamp	White
Stop lamp	Amber
Direction indicators	Red
Rear reflectors	White
Number plate lamp	Red

It is important to note that some motorcycles have daytime running lights (DRLs) fitted. These are designed to come on whenever the engine is running. They do not use the low beam headlights and must deactivate when normal headlights come on.

Why are DLRs fitted to motorcycles? _____

Battery

The main function of the battery is to provide power for the starter motor. Construction is similar to that of a light or heavy vehicle battery and they are generally of the lead acid type. Most motorcycles use either 6 V or 12 V electrical systems.

Visually check the battery fluid levels, if they are low top them up with distilled water. Check the battery terminals for security. The motorcycle uses a chassis earth return by using the frame as the negative return to the battery.

Alternator

When the engine is running the alternator charges the battery. It also provides the electrical current for the lighting, horn and ignition system that produces the spark for the engine.

TECHNICAL DATA

When working on motorcycles it is always important to refer to the manufacturer's technical data. Technical data are available in different forms. List THREE of these:

1 _____

2 _____

3 _____

Complete the following table stating the type of technical data you should analyse to check the component is functioning correctly.

Component	Type of technical information
Brake/clutch fluids, engine oil	_____
Tyres	_____
Nuts and bolts	_____

Look at a moped and a 1000 cc motorcycle in your garage and complete the following tables to understand how the two vehicle types differ.

Moped model:

Type of technical information	Technical information
Engine capacity	
Power output	
Service intervals	
Fluid types and capacities	
Ideal tyre pressures	

1000cc motorcycle model:

Type of technical information	Technical information
Engine capacity	
Power output	
Service intervals	
Fluid types and capacities	
Ideal tyre pressures	

Multiple choice questions

Choose the correct answer from a), b) or c) and place a tick [✓] after your answer.

1 **How many brake reservoirs do motorcycles normally have?**

   a) Four [ ]

   b) Two [ ]

   c) Three [ ]

2 **A moped is a motorcycle that weighs:**

   a) Less than 250 kg [ ]

   b) More than 250 kg [ ]

   c) Less than 410 kg [ ]

3 **DRL is an abbreviation for:**

   a) Daytime Road Lights [ ]

   b) Daytime Rear Lights [ ]

   c) Daytime Running Lights [ ]

4 **A lead acid battery needs to be topped up with:**

   a) Sulphuric acid [ ]

   b) Distilled water [ ]

   c) Tap water [ ]

5 **Why is it important not to spill brake fluid onto painted surfaces?**

   a) Brake fluid is expensive [ ]

   b) It could damage the paintwork [ ]

   c) Brake fluid is hard to remove [ ]

GLOSSARY

4WD Four-wheel drive

Ackerman linkage Form of steering arranged to give true rolling motion round corners

After sales (service) department Section of a garage dealing with routine servicing, diagnosis and repair of faults and often MOT tests

Air gap The gap between the electrodes of a spark plug or reluctor and pick-up

Alternating current (AC) Electricity that moves in two directions

Alternator Charges the battery and provides electrical current for the motorcycle electrical systems

Amplifier A device used to increase the electrical signal in an electronic ignition system

Amp The unit of measurement for electrical current flow

Antifreeze Added to water to lower freezing point and protect the engine from corrosion

Aspect ratio The ratio between the height and width of a tyre (expressed as a percentage)

AWD All-wheel drive

Balancing Correcting any unbalance of wheel and tyre assemblies

Battery Used to power all of the electrical components located within the fuel system

Beam axle Rigid axle

Biofuel Fuel that is obtained from growing crops

Boundary A term applied where the film of lubricant is applied by splash and mist with the possibility of some metal-to-metal contact

Brake fade Loss of friction and braking force due to overheating brakes

Bump Upward movement of suspension

Caliper Housing for the piston(s) which moves the brake pads to contact the rotating disc

Camshaft Rotates in the engine and opens the valves at the correct time

Carbon dioxide (CO_2) Greenhouse gas that contributes to global warming

Carbon footprint The total emissions caused by an organization, event, product or person

Carbon monoxide (CO) Colourless, odourless, poisonous to human and animal life

Carburettor A mechanical device for mixing fuel with air

Catalytic converter Converts harmful gases into less harmful gases

Chains Used to transmit power to the motor cycle wheels

Chamois Automotive Drying cloth that is safe on acrylic, lacquer, enamel, and polyurethane paints and clear-coats

Clutch plate Component splined to the gearbox input shaft (also known as centre plate or driven plate)

Clutch slip Sometimes referred to as feathering the clutch. When the clutch plate slips against the flywheel when the driver applies and releases the clutch pedal

Clutch Provides a temporary position of neutral and enables a smooth take-up of drive

Coil An ignition coil is able to transform battery voltage to the very high voltages required for a spark to cross the electrode gap of a spark plug

Communication Listening, reading, speaking and writing

Conduction When heat passes through solid materials, mainly metals

Constant velocity joint Driveshaft joint that allows steering and suspension movement when transmitting drive to the wheels

Contact breaker A mechanical switching device used in older ignition systems

Contract of employment An agreement between an employer and an employee

Convection When heat is carried by moving liquid or gas in an upwards direction

Coolant sensor Senses the temperature of the vehicle's coolant

Cooling fan Helps to cool the radiator and maintain the coolant temperature under extreme conditions

COSHH Control of Substances Hazardous to Health

Crankshaft position sensor (CPS) Designed to signal the engine management system when the engine is in the correct position to run

Crankshaft Bolted into the lower part of the engine block. It converts linear motion into rotational motion

Cross ply A tyre in which the plies are placed diagonally across each other at an angle of approximately 30 to 40 degrees

Current The flow of electricity

Curriculum Vitae A document containing details of someone's course of life

Cutting compound A paste used for removing scratches, blending in new paintwork and removing paint over-spray

Damper A device which dampens the oscillations or vibrations of the road springs

Dead axle Non-driven axle (usually on the rear)

Direct current (DC) Electricity which moves only in one direction

Disc brake A brake in which friction pads grip a rotating disc in order to slow the vehicle down

Discrimination Prejudicial treatment of an individual based on their race, gender or religion

Distributor A mechanically driven component, which distributes high voltage to the spark plugs

DIS Distributor less ignition system

Drag link Connects drop arm to first steering arm

Drive belt/serpentine belt Commonly referred to as the fan belt. This can be used to drive the water pump, alternator, power steering pump and some auxiliary units

Drop arm Connects steering box to drag link of steering system

Drum brake A brake in which curved shoes press on the inside of a metal drum to produce friction in order to slow the vehicle down

Dry sump Oil is supplied and returned to a separate oil tank away from the engine

Dwell The period of time when electricity is passed into the coil (typically 3–6 milliseconds)

Electric vehicle Totally reliant on an electric motor for propulsion

Electrolyte A mixture of distilled water and sulphuric acid which is used in a battery

Engine control unit (ECU) The electronic brain of the vehicle where all sensors send inputs to monitor the engine efficiency

Engine laquer A fast-drying clear acrylic varnish which enhances the appearance of the engine and its components. The varnish protects the vehicle from moisture and corrosion

Exhaust gas recirculation valve (EGR) Controls the amount of exhaust gasses that are recirculated back into the engine

Expansion tank Fitted to some cooling systems as a reservoir

Ferrous A metal which contains iron, making it magnetic

Final drive Takes drive from the gearbox to the driven wheels

Firing order This is the order that combustion takes place in multi-cylinder engines

Flywheel Bolted to the crankshaft, transmits engine torque to the clutch plate

Four stroke The operating cycles for complete and full combustion in the Otto cycle

Franchised dealer This is a garage dealership linked to a vehicle manufacturer, can be known as a main dealer

Fuel filter Prevents any dirt in the fuel reaching the carburettor/injectors and contaminating the fuel system

Full-fluid film A film of lubricant that is sufficiently 'thick' so that no metal-to-metal contact takes place

FWD Front-wheel drive

Gearbox input shaft First shaft to turn in the gearbox, connected to the clutch plate

Gearbox output shaft Takes the drive to the final drive assembly

Gearbox A major unit which multiplies the engine torque (turning force) and provides a means of reversing the vehicle, as well as a permanent neutral

Glow plug Heats the air in the combustion chamber to aid cold starting

Greenhouse gases The harmful gases and emissions released into the atmosphere.

Handlebars Used for steering a motorcycle or moped. This is also a typical mounting place for the vehicle's controls

HASAWA Health and Safety at Work Act (1974)

Hazard Potential to cause harm or injury

Heavy Goods Vehicles (HGV) or Large Goods Vehicles (LGV) Any truck weighing over 3500 kilograms according to European Union regulations

High tension circuit A circuit that carries anything up to 40 000 volts

High tension leads Carry high voltage from the distributor to the spark plugs

HSE Health and Safety Executive

Hybrid Vehicle using a combination of power sources such as conventional engine and electric motors

Hydrocarbon (HC) Causes respiratory problems, liver damage and cancers

Hydrodynamic lubrication Using the natural movement of the oil 'wedge' to separate the surfaces of highly loaded bearings when shafts rotate

Hydrometer Equipment used to test the specific gravity of a liquid

Hygroscopic An ability to absorb moisture from the atmosphere

Idler arm Similar to the drop arm but having a guiding function only

IFS Independent front suspension

In-line An engine layout where the cylinders run from front to back

Injection pump Increases the pressure of the diesel sufficiently for injection

Injector Injects diesel into the combustion chamber under very high pressure

IRS Independent rear suspension

Job description Document that give specific details of a job role

Lambda sensor Measures the oxygen content in the exhaust gases

Leading shoe One of the shoes in a brake drum assembly which pivots outwards into the drum first

Live axle Driven axle

Load index How much weight a tyre will safely carry

Low pressure pump or lift pump Used to pressurize the fuel injection system fuel lines/pipes used to transport safely the fuel to and from the tank

Low tension circuit A circuit that carries normal battery voltage

Master cylinder The largest cylinder in the hydraulic circuit which pressurizes the fluid

Moped A motorcycle that weighs less than 250 kg and has a maximum design speed not greater than 30 mph

Motorcycle A vehicle having less than four wheels and weighing less than 410kg unladen

Multi-point An injection system in which each cylinder has its own injector. Only air enters the inlet manifold. The injectors are situated in the inlet manifold close to the valve ports

Multigrade Oil which meets the viscosity requirements of several different single-grade oils

Ohm The unit of measurement for electrical resistance

Organizational structure Plan showing the relationship between job roles and lines of authority (who is responsible to who)

Oxides of nitrogen (NO_x) Can cause respiratory conditions, smog and acid rain

Panhard rod A rod mounted between the body or chassis and the axle, to control the lateral (sideways) movement of the axle

Parts department Section of a dealership selling spare parts to trade and retail customers

Passenger Carrying Vehicles (PCV)

Pinion Gear wheel which engages with a rack

Pitch Forward and backward rocking motion of the vehicle

Plies Increase the strength of the tyre

PM soot particles Cause respiratory problems and cancers

PPE Personal protective equipment

Pressure cap Maintains the correct operating pressure of the cooling system

Public Service Vehicles (PSV) A vehicle that carries more than eight passengers for a fare e.g. bus or coach

Rack Toothed bar

Radial A tyre in which the plies are placed at right angles to the rim

Radiation When heat is given off into the air, from the surface of the object

Radiator Coolant flows through it and transfers heat from the engine to the surrounding air

Rebound Downward movement of suspension

Remuneration Employee's compensation for work done for the employer. This is usually a wage. It can also include complementary benefits like free health insurance.

Risk Likelihood or chance of harm being caused

Roll Sideways sway or 'leaning outwards' of a vehicle on corners

Rubber dressing Cleans and restore tyres, rubber trims and interior plastics to an 'as new' appearance

RWD Rear-wheel drive

Secondary brake The emergency brake. This is not powered by hydraulics and is independent of the service brakes used to slow and stop vehicles

Self-servo action Self-energizing effect which helps to multiply the braking force when the brake shoe contacts the drum

Service brake The primary braking system. This brake is typically operated by foot and is mechanically separated from the parking brake or emergency braking system.

Silencer Reduces engine noise to an acceptable level

Single point A single injector system which sprays fuel for all cylinders into the air at one place, usually by the throttle body in the inlet manifold

Soot particles Cause respiratory problems and cancers

Speed rating Maximum safe tyre speed

Spring A suspension device designed to absorb road shocks, i.e. a spring

Sprocket A profiled wheel with teeth

Sprung weight All of the vehicle's mass that is above the spring

Steering box Changes rotary movement into linear movement

Stoichiometric Chemically correct ratio of fuel and air required for complete combustion

Sulphur dioxide (SO$_2$) Can cause pollution and acid rain

SWL Safe Working Load. The maximum weight that can be lifted or supported

Tank Used to safely store fuel for the engine

Teamwork This means cooperating with and caring about other workers

Terms of contract The employee's employment rights, responsibilities and duties set out in the contract of employment

Thermostat Temperature sensitive valve that controls coolant flow to the radiator

Tie rod Connecting rod or bar, usually under tension

Toe in or toe out Inward or outward inclination of the leading edge of the front wheels

Torque Turning force measured in Newton metres (Nm)

Track rod Bar connecting the steering arms

Trade union An organization made up of members of workers. One of a trade union's main aims is to protect and advance the interests of its members in the workplace

Trailing shoe The shoe in a brake drum assembly which is forced away from the drum by its rotation

Transistor A semi-conductor which can be used to switch electronic circuits and also amplify voltage

Transverse An engine layout where the cylinders are positioned from side to side (across the vehicle)

Tube Fitted inside the tyre to retain air

Un-sprung weight All of the vehicle's mass that is below the spring

Universal joint Propeller shaft joint that allows small angular changes

Valve timing The correct time when the valves open and close

Viscosity The resistance to flow or 'thickness' of a liquid. It can also be described as its resistance to shear

Viscosity index A number which indicates how the viscosity of a liquid changes with temperature

Volt The unit of measurement for electrical pressure

Water pump Circulates coolant around the engine

Watt The unit of electrical power

Well base rim A rim with a centre channel which enables easy removal and refitting of the tyre

Wet sump or reservoir Oil is returned from the engine by gravity and collected in a sump below the engine